Motherhood,
the Elephant in the Laboratory

Motherhood,
the Elephant in the Laboratory

Women Scientists Speak Out

EDITED BY

Emily Monosson

ILR Press
an imprint of
Cornell University Press
Ithaca and London

First published 2008 by Cornell University Press

Printed in the United States of America

Library of Congress Cataloging-in-Publication Data

Motherhood, the elephant in the laboratory : women scientists speak
out / edited by Emily Monosson.
    p. cm.
  Includes bibliographical references.
  ISBN 978-0-8014-4664-1 (cloth : alk. paper)
  1. Working mothers—United States—Biography. 2. Women
scientists—United States—Biography. 3. Women scientists—
Family relationships—United States. 4. Motherhood—United
States. 5. Work and family—United States. I. Monosson, Emily.
II. Title.

  HQ759.48.M68 2008
  306.3'6—dc22

Cornell University Press strives to use environmentally responsible
suppliers and materials to the fullest extent possible in the
publishing of its books. Such materials include vegetable-based,
low-VOC inks and acid-free papers that are recycled, totally
chlorine-free, or partly composed of nonwood fibers. For further
 information, visit our website at www.cornellpress.cornell.edu.

Cloth printing   10 9 8 7 6 5 4 3 2 1

To all the women in whose footsteps we've followed,
and to those who choose to follow in ours

Contents

Acknowledgments

This book evolved from a single desperate e-mail sent to a Listserv for current and former fellows of the American Association for the Advancement of Science, and I am indebted to those women (and few men) who responded, creating the community from which this project grew. A second round of e-mails resulted in a larger community of women willing to share their experiences so that others might benefit, and I am grateful to all of them; without their essays there would have been no book. I also acknowledge Fran Benson at Cornell University Press, who was willing to take a second look and who was very patient with my impatience and persistence; and Cameron Cooper and Candace Akins, also at Cornell University Press, for helping to whip this collection into shape. Over the past year and a half I have recruited many friends, family members, neighbors, and colleagues to read and comment, draft after draft. In particular I thank Penny Shockett, Marla McIntosh, Dori Ostermiller, my kids—Sam and Sophie—and most of all, my husband, Ben Letcher, who was willing to drop whatever he was doing (usually) whenever I asked, "Can I read this to you?" Ben has supported and encouraged me throughout this project and throughout my life with him.

Introduction

Initiating the Discussion

"Most of us thought we would work and have kids, at least that was what we were brought up thinking we would do—no problem. But really we were kind of duped. None of us realized how hard it is."[1]

This quote hit home. I am a split personality, the product of my mother—whose job it was to keep the house, raise the kids, and support my father—and my father, who loved his work and held in highest esteem the university faculty who taught him about science, math, and business. Although I strive to be like my mother, I aspired to become a scientist ever since the day my father, with boyish glee tempered by parental caution, dumped a mixture of chemicals from his old chemistry set into a hole in the ground and we watched them hiss, bubble, and fade into the earth.

When I received my doctorate in toxicology from Cornell University, my father tacked up the framed photo of me shaking hands with Frank Rhodes, then president of the college, on his office wall. It was the only

1. Cathie Watson-Short, *New York Times* Quote of the Day, in Eduardo Porter, "Stretched to the Limit, Women Stall March to Work," *New York Times*, March 2, 2006.

photo he'd kept in his office of any one of his four grown girls. So on the day I announced that I was moving from my research position in Rhode Island to an uncertain future in North Carolina, accompanying my soon-to-be fiancé as he pursued his PhD, my father called. "Lemme ask you a question," he said. "What about your research?" His fear that I might throw it all away, for a man he'd not yet met, was evident. Yet several years after that, while happily married and working as a research associate, when I announced my first pregnancy, he expressed nothing but joy. Perhaps by then he believed his youngest could do it all. But as I transformed from a full-time laboratory researcher to a homebound scientist surrounded by piles of reprints, half-eaten finger foods, and balled-up diapers, I found myself presenting two not entirely realistic selves to my father: one, the fully dedicated scientist and ideal worker; the other, the ideal mother whose first priority was her babies.

For years I'd wondered what was wrong with me. Since I'd decided that I would work only during school hours while the kids were young, was I not a dedicated scientist? Guiltily I wondered if I'd set a poor example for young women in science. I grew up in the 1970s when women fought for equal rights. When my father, who constantly encouraged us to pursue our passions in life, dared one evening to acknowledge to his wife and girls (four sisters) that he was reluctant to hire a young woman for a high-level position with his company (a company that one of my sisters now heads) for fear that she might get pregnant and leave, the five of us pounced. It wasn't pretty.

Thirty years later, had I become one of those women? After reading Watson-Short's quote I realized I wasn't alone in making such difficult choices. Empowered by that knowledge, I sat down in my home office and typed a short note to the Listserv for former American Association for the Advancement of Science Fellows, or AAAS Fellows, one of my links to scientists from around the country, attached the *New York Times* article along with the quote, closed my eyes, and hit Send. Outing myself by broadcasting the article was an act of desperation. I was admitting to an elite group of scientists that I am a mother who struggles to succeed as a scientist and a scientist who finds it difficult to be an ideal mother. I wanted to know I was not alone.

Responses were immediate, enthusiastic, and emotional. For many women, this was the first time anyone had asked that they share their experiences without being judged for their choices. Though these women responded with passion, many wished to keep their responses anonymous.

Some were uncomfortable discussing family and work practices on a forum for science professionals. On the Internet it is easy to assume the persona of a full-time ideal worker. Some respondents were afraid that if they discussed difficulties of combining career and family, they'd be charged with whining. Many, however, felt that by posting their comments to the list, they might encourage others to come forward, initiating a broader discussion about combining motherhood and a career in science:

> The push to get more women in science and engineering has ignored the elephant in the room—motherhood. (Denise DeLuca, PE, Outreach Director, The Biomimicry Institute)

> I really appreciate your raising this issue, despite everyone's reluctance to discuss it openly. (Rachel S., PhD)

> In the final analysis, every woman finds her own way. It's just good to know that none of us is alone. (Frieda S., PhD)

Scientists with families, particularly women with young children, find it difficult to achieve a balance between work and family in these highly competitive, often male-dominated fields. And it is not just the sciences. The media, academic journals, and libraries abound with articles and books detailing the struggles and difficult decisions faced by working parents (though primarily women) in a range of professions from engineering to law to academics.[2]

Although about half of the undergraduate and over 40 percent of graduate degree recipients in science and engineering are women, in 2003 they represented only 27 percent of all employed doctoral-level scientists, reaching parity with men in just a handful of science occupations such as psychology (as psychologists) and postsecondary teaching for health and

---

2. Peter Meiksins and Peter Whalley, *Putting Work in Its Place* (Ithaca: Cornell University Press, 2002); Leslie Perlow, *Finding Time: How Corporations, Individuals, and Families Can Benefit from New Work Practices* (Ithaca: Cornell University Press, 1997); C. Taylor, "Scientists as Parents," *ScienceCareers: The American Association for the Advancement of Science* (January 2004), http://sciencecareers.sciencemag.org/career_development/previous_issues/articles/2800/scientists_as_parents_feature_index; Lucille Louis, "The X-gals Alliance," *ChronicleCareers: The Chronicle of Higher Education* (October 2006), http://chronicle.com/jobs/news/2006/10/2006 100201c/careers.html; Robert Drago, "Harvard and the Academic Glass Ceiling," *ChronicleCareers: The Chronicle of Higher Education* (March 2007), http://chronicle.com/jobs/news/2007/03/2007 032701c/careers.html.

related sciences.[3] In the category of "contingent" faculty members, those who work part-time or full-time as non-tenure-track faculty, the proportion of women working as contingent faculty exceeds the proportion of men.[4] These data have not gone unnoticed, and one needn't look far to find programs, studies, and books aimed at solving the case of the vanishing woman scientist, a phenomenon sometimes referred to as "the leaky pipeline,"[5] particularly in what has been traditionally considered the pinnacle of scientific success, academia.[6]

But if women really are leaving the sciences, where are they going? We're talking about thousands of women. Do they seek alternative paths? If so, do they continue to contribute to the scientific community or to science in some way? If they leave, what impact does this have on science and society? Though these critical questions have been addressed by two recent National Academy of Science publications they provide few answers to the question, "Where are they going?"[7]

This book contains essays written by thirty-four mother-scientists whose stories provide insight into the choices they have made to create balance in their lives. Contributors to this book work part-time or full-time, opt out, and opt back in. They've become entrepreneurs, they job-share, and they volunteer. They work in academia, industry, consulting, state and federal government, and on their own. Some of these women who have chosen to stray from the straight and narrow road paved by mentors, ad-

3. National Science Foundation, Division of Science Resources Statistics, *Characteristics of Doctoral Scientists and Engineers in the United States* Survey of Doctoral Recipients (Arlington, Va.: National Science Foundation, 2003), table 14, http://www.nsf.gov/statistics/nsf06320/tables.htm.

4. E. Ivey, C. Weng, and C. Vahadji, *Gender Differences among Contingent Faculty: A Literature Review*, Final Report, The Association for Women in Science, 2005, http://www.awis.org/pubs/sloanreport.pdf.

5. Yu Xie and Kimberlee Shauman, *Women in Science: Career Processes and Outcomes* (Cambridge, Mass.: Harvard University Press, 2003). The term "leaky pipeline" is discussed by Xie and Shauman in the introduction to their book, pages 6–9.

6. Xie and Shauman, *Women in Science;* National Academy of Sciences, National Academy of Engineering, Institute of Medicine of the National Academies, *Beyond Bias and Barriers: Fulfilling the Potential of Women in Academic Sciences and Engineering* (Washington, DC: National Academies Press, 2006); National Science Foundation, Division of Science Resource Statistics, *Gender Differences in the Careers of Academics, Scientists and Engineers* (Washington, DC: National Science Foundation, 2004); *ADVANCE: Increasing the Participation and Advancement of Women in Academic Science and Engineering Careers* (Arlington, VA: National Science Foundation), http://www.nsf.gov/funding/pgm_summ.jsp?pims_id=5383.

7. National Academy of Sciences, *Beyond Bias and Barriers;* and *Rising above the Gathering Storm: Energizing and Employing America for a Brighter Economic Future* (Washington, DC: National Academies Press, 2007).

visors, and scientists before them by working part-time, or who no longer coax data from the bench or the field, have a sense that they have become an invisible, underutilized, and misunderstood workforce. They often feel marginalized when they attempt to return or interact with the more traditional workforce. Their feelings are summed up by M. T., who has worked as an editor, research associate, and volunteer:

> I find myself constantly rehearsing and drafting what I will say to people I meet at meetings and in professional settings about my unpaid research situation and all the volunteer work I do to promote programs for government agencies, professional societies, and education. (M. T., PhD)

M. T. is not alone. There are others, women in particular, who seek alternatives and who contribute to the sciences in nontraditional ways, their choices driven in large part by a desire for an acceptable work-life balance; they could use support and encouragement from the larger scientific community. As one graduate advisor responded to the original e-mail:

> The graduate students in our department frequently complain about not being educated about career options outside of traditional academic careers. When, as graduate studies chair, I talked one-on-one with female students trying to figure out how to make life work (this happened a lot—I always wondered whether the male graduate studies chairs were approached about this as well), I tell them about women who are tenured, or who teach high school or who work part-time as examples of different ways to have successful lives when children arrive after PhDs. . . . I also talk to students about not letting themselves define their goals and success by their advisor's (or their perception of their advisor's) ideas of success. (Libby Marschall, PhD, Department of Evolution, Ecology, and Organismal Biology, Ohio State University)

My motivation for compiling this book was to highlight the accomplishments, challenges, and choices made by women scientists as they combine motherhood and career. I've included essays written by women who have chosen routes outside the mainstream as well as those who have followed traditional career tracks in academia or as government researchers. Essays are organized chronologically by date of last degree conferred, and contributors range from women who received their PhDs in the 1970s to those still in graduate school. Because of the variety of experiences reported by these women, organizing essays by work sector (academia, industry, government) or by time spent in the workforce (full-time, part-time, opting out, and opting in) was too limiting. In the end, a chronological organization, tracking the interaction of science and motherhood across a span of time in which drastic changes in both science and women's rights have occurred, made the most sense.

In all cases, when there is family involved, there is a story to tell. Sharing these stories serves others by reassuring, encouraging, or cautioning them as they seek the balance that works for them. My goals for this book are twofold: to initiate discussion on redefining the concept of "career" scientist and to examine the many different ways in which women have managed to combine motherhood with their science careers. Writes Rachel, another early e-mail respondent:

> I can only hope that by continuing to have the discussion, that ultimately policies and society will change to become more egalitarian and family friendly. (Rachel S., PhD)

Defining the Boundaries

The first time I talked to a group about this book, I was asked how I had defined "scientist." It was a good question, and I did not have a satisfactory answer. While gathering essays, I'd inadvertently narrowed my definition of a scientist to someone who had earned a PhD in the natural and physical sciences (though I had let a few engineers and social scientists slip in as well). I thought this would provide a clear demarcation. Then one woman asked if a master's degree with ten years of experience qualified. Another, who has a PhD but now teaches high school science, wondered if she still counted.

"Of course," I'd answered to both, based on my (perhaps self-serving) belief that the definition of a scientist includes much more than the traditional sum of her degrees, grants, and publications. When I think about the many scientists I know, science is not only their profession but a way of thinking about the world, a way of life. Scientists find joy in science. We ask questions, seek answers, are curious. If we did not love our work, the four, five, six, or more years of graduate school (often during prime childbearing years) would be a far too painful sacrifice. I've yet to hear a scientist describe her (or his) work as "just a job."

Many of us mothers who leave the mainstream, or leak from the pipeline, will do whatever it takes to nurture and grow our scientist selves.

But are the women who have pursued alternatives to careers in academia or as primary investigators of a research laboratory, seeking work-life balance, still considered scientists by the larger scientific community? Some would say no. Once again I turned to the AAAS Listserv this time asking (1) how would you define scientist? and (2) how would you characterize success in science? In response to the former, I received the following e-mail from Ravi Sawhney, an orthodontist and cell biologist, now working on science policy at the National Institutes of Health, which despite my own broad definition, resonated with the more traditional part of me. Wrote Ravi:

1) A scientist is someone who spends a significant portion of their time,

2) using the scientific method,

3) to answer questions, test hypotheses, or build models that lead to predictions,

4) in order to further the human understanding of the workings of nature.

Point 1 is because everyone dabbles in science whether trying to figure out how to lure a mate, raise kids, or just how much Weed & Feed you need to kill the damned dandelions. "Everyone" isn't a scientist though.

Point 2 is because I think you actually have to be practicing science. Teaching science is extremely important to the scientific enterprise, and teachers are a valuable return on our investment in research, but teaching science doesn't make one a scientist . . . any more than someone teaching art makes someone an artist in itself. An art teacher may have a much more significant impact on the world than an individual artist. It isn't a value judgment, just my definition. It also implies that just observing and describing, or using high tech gadgetry, or thinking a lot, etc., doesn't make one a scientist.

Point 3 is what science is. Everyone has a different definition.

Point 4 is because I think you have to actually put your data out there to call yourself a scientist. I think a scientist actually has to be advancing scientific understanding.

Ravi concluded by adding,

I was trained as a scientist. As a Health Science Policy Analyst at the NIH, I think about science all the time; I try to advance it; I try to help the world understand how important it is. I field tough questions at dinner parties. But, as much as I hate to admit it . . . gulp . . . I am no longer a scientist.

Thanks for making me face that brutal reality. Dear God, when did I go astray?[8]

Ravi's definition was both thorough and, given my current status, somewhat depressing. Although some part of me agreed with Ravi, my inner scientist, fully aware of her own bias, begged me to keep searching.

I discovered that Merriam-Webster provides a more liberal definition, describing a scientist as one who is "learned in science and especially natural science," and defines science as "knowledge or a system of knowledge covering general truths or the operation of general laws especially as obtained and tested through scientific method."

On the basis of my own experiences and those of friends, colleagues, and those who have contributed to this book, either by writing essays or by participating in the first few rounds of e-mail, I would suggest a combination of the two definitions. I'm not sure "being learned," which these days may imply a PhD (or in some cases an MS followed by independent research), is enough. I believe that part of being a scientist, as Ravi describes, is advancing scientific information, using the knowledge and the scientific method, whether by designing experiments in a research laboratory, developing an ecology field trip for high school seniors, preparing an analysis based on literature review, or educating communities about groundwater issues.

I think it is important to add here a brief note about the term "career," which also has several connotations. In her analysis of women and work, discussed in next section, Claudia Goldin acknowledges that "career" is difficult to define and that "in common parlance, it means a success that is not ephemeral." In need of a technical definition for her analysis, however, she then more narrowly defines a woman with a career as "earning more than a college graduate man whose income is well below that of the median man (but about equal to the median of the female earnings distribution) for several consecutive years."[9] Such a definition would likely exclude several contributors to this volume of essays (present company included).

---

8. Ravi later wrote back that after he shared his definition, it was "roundly rejected—by artists who felt it cruel to say that an actress having to wait tables to make ends meet is no longer an actress, and by researchers who felt it unfair to say that someone who lost their NIH grant and thus their lab is no longer a scientist. Perhaps," added Ravi, "the intention to do science may be more important than the actual doing of science in defining a scientist."

9. Claudia Goldin, "The Long Road to the Fast Track: Career and Family," *ANNALS AAPSS* 596 (2004): 20–35, at 31.

Yet, as Peter Meiksins and Peter Whalley, authors of *Putting Work in Its Place*, write in reference to careers:

> [B]eing serious about one's work is supposed to mean a full-time, indeed an extended time commitment. . . . This is what is traditionally meant by having a career. . . . Careers not jobs, are what help shape identities, give form to a work life, and gain public recognition. . . . Professional women with children [referring to those who choose flexible and part-time work options], in particular, have to resist the assumption that they have settled for the mommy track, a less demanding form of work, not really a career, just a job (although the man or woman in the next cubicle or office may be doing similar work but be on the fast track to the top).[10]

In one form or another, the contributors to this book have chosen to dedicate their lives to a career in science, whether it is teaching, research, or policy.

The second issue, once we decide who still belongs to the science club, is success. I've added success, because "success in science" is a concept that appears in reports about the leaky pipeline or the vanishing woman scientist. The perception is that not only are women leaving the sciences but also that many women are not achieving a certain standard of success.

Because in certain disciplines academia remains the ultimate and most desirable outcome for scientists, some scientists who leave express guilt and a sense that they have failed their advisors, or that they are letting other women down, perhaps even setting a poor example. Additionally, those who choose careers they consider more amenable to raising children or who take time from their full-time positions fear that discussing the impact of motherhood on their careers will weaken their professional standing and future career options.

Reading the contributed essays and observing the careers of scientists both inside and outside academia, I would suggest that a broader and more inclusive definition of success (beyond attainment of tenure) in science might lead to a more inclusive and perhaps more welcoming scientific community, one that does not discourage but encourages the participation of all kinds of scientists in all kinds of roles. To do otherwise would be to label as failures those scientists who leave the academic pipeline; who are lecturers, adjuncts, or high school teachers; who choose the position of re-

10. Meiksins and Whalley, *Putting Work in Its Place*, 35–36.

search associate rather than primary investigator; or who choose policy or writing. Should success in science be measured purely by the type and size of a grant, the number of publications, and the number of graduate students trained? Or is there a place for a broader definition of success that values contributions to science that cannot be measured with dollar signs or quantities of goods?

The following e-mails about success suggest there is room for more than one definition:

> Some weeks ago, a colleague and I talked about how we were all brainwashed with the "publish or perish" rule, and we were warned that we must have grants in highly competitive settings in order to succeed in science careers. Now, she and I and many others have found very productive and interesting careers by ignoring that "old school" advice. (Alexandra S. Fairfield, PhD, National Institutes of Health, retired)

> We consider a trainee a success even if they are in a policy or administrative position that deals with Science. Our thought is that, like the AAAS fellowship acknowledges, we need scientists in administration and policy to help translate scientific discovery into informed policy decisions. . . . [W]e use a very broad definition of success. (L. K., PhD, former AAAS Science and Diplomacy Fellow, Fogarty International Center, NIH)

> Yet another way to think about [success] is to consider what defines a successful scientific community, rather than what defines a successful individual scientist. In my own opinion, a successful scientific community requires talented researchers, science teachers, science writers, science advocates, and people in many other science-related areas. (Rebecca Farkas, PhD, AAAS Science Policy Fellow)

I believe there is room for a definition of success that is not limited to appointments at respected universities and laboratories or prestigious grants that support large laboratories. In my own field of environmental sciences, at least two large movements were initiated by inspiring women who worked outside academia and who did not run large laboratories—Rachel Carson, credited with initiating the environmental movement, and Theo Colborn, who helped draw attention to the consequences of endocrine-disrupting chemicals in the environment. These women observed and synthesized the work of many others and drew insightful conclusions.

As Rebecca noted in her quote above, a successful scientific community requires a diversity of members. Scientific advances require those who discover the impact of ocean currents on global temperature and novel applications for nanomaterials and those who educate and inspire the next generation of scientists. Scientific advances also require those who inform politicians and lobby for the funds to support these scientists. Application of scientific advances requires those who inform policymakers and the public about the importance of the risks and benefits of new technologies. And women with children populate all these niches—some choosing one over another to accommodate family.

The Elephant

Women are an integral part of the larger scientific community. According to the National Science Foundation (NSF), there are approximately one hundred thousand women doctoral degree holders in the United States employed in the sciences,[11] but employment figures, particularly in academia, suggest women are leaving the sciences in droves.

In the fall of 2006, the National Academy of Sciences released its highly quoted report *Beyond Bias and Barriers*, the goal of which was to "develop specific recommendations on how to make the fullest possible use of a large source of our nation's talent: women in academic science and engineering."[12] That they were compelled to state up front not only that women "have the ability and the drive to succeed" but that the lower representation of women in the highest reaches of academic math and sciences can't be accounted for by any "significant biological differences between men and women in performing science and mathematics" would have been laughable to the hundred thousand women scientists had it not been for the comments made in a speech the previous winter by the now former president of Harvard University, Lawrence Summers. Addressing the National Bureau of Economic Research, Summers suggested that the under-representation of women in science had both a biological and social basis.[13]

11. National Science Foundation, Division of Science Resources Statistics, *Characteristics of Doctoral Scientists and Engineers*, tables 26–29, http://www.nsf.gov/statistics/nsf06320/tables.htm
12. National Academy of Sciences, *Beyond Bias and Barriers*.
13. Lawrence H. Summers, "Remarks at NBER Conference on Diversifying the Science and Engineering Workforce," Cambridge, Mass., January 14, 2005, http://www.president.harvard.edu/speeches/2005/nber.html.

His comments created a major backlash,[14] leading to his eventual resignation and prompting social scientists Stephen Ceci and Wendy Williams to solicit "evidence-based" essays debating gender differences in cognition.[15] Writing about Summers's comment, Ceci and Williams remark:

> Coming as it did from the gatekeeper of one of the world's great institutions of higher learning, the insinuation of biologically based differences in cognition, coupled with an accusation that advocates of greater equity for females in science were grasping at weak socialization explanations, was radioactive. . . .[16]

Although Ceci and Willams invite readers to decide for themselves why more women aren't in science, they write in their conclusion, "Sex differences appear to be neither as unambiguous as earlier researchers suggested nor as insubstantial as some current critics claim. Sex differences in career choices are definitely not inevitable as the past 30 years have documented a sea change in the gender makeup of various fields."[17]

These days, almost 45 percent of all science and engineering PhD recipients are women.[18] There is no doubt that women are attracted to and can succeed in the sciences. Yet many leave, while others favor certain less visible and lower-paying sectors, including educational institutions other than four-year colleges and universities, private not-for profit organizations, and self-employment.[19] In academia, compared with male faculty, a greater proportion of women faculty work in non-tenure-track positions as lecturers and adjuncts.[20] More female than male PhD recipients report

14. Marcella Bombardieri, "Summers' Remarks on Women Draw Fire," *Boston Globe*, January 17, 2005, http://www.boston.com/news/local/articles/2005/01/17/summers_remarks_on_women_draw_fire; Letter from President Summers on women and science, January 19, 2005, http://www.president.harvard.edu/speeches/2005/womensci.html; Women in Science and Education Leadership Institute (WISELI), University of Wisconsin, Madison, has a website devoted to his comments and responses to his comments: Responses to Lawrence Summers on Women in Science, http://wiseli.engr.wisc.edu/news/Summers.htm.

15. Stephen Ceci and Wendy M. Williams, eds., *Why Aren't More Women in Science?* (Washington, DC: American Psychological Association, 2007).

16. Ibid., 8.

17. Ibid., 223–24.

18. National Science Foundation, Division of Science Resources Statistics, *Characteristics of Doctoral Scientists and Engineers* (2005), figure F-1, http://www.nsf.gov/statistics/wmpd/figf-1.htm.

19. National Science Foundation, Division of Science Resources Statistics, *Characteristics of Doctoral Scientists and Engineers* (2003), table H-11, table H-12, table H-33, http://www.nsf.gov/statistics/nsf06320/tables.htm.

20. Ivey, Weng, and Vahadji, *Gender Differences*, 14.

that they are either employed part-time or not employed and not seeking work.[21] Why?

One answer is family. Of those working part-time, half indicated that they chose part-time work to accommodate family.[22] Although numbers aren't available for full-time workers who chose nonacademic careers because of family responsibilities, evidence suggests that for women family considerations weigh heavily. For those who are tenured or in tenure-track positions, the National Academy of Sciences found that women "consistently ranked working conditions, family, and job location higher than men among their reasons for changing jobs." Further, the study found that "women are 40% more likely than men to exit the tenure track for an adjunct position,"[23] and although reasons for such changes were not explicit, a growing number of journal articles and news stories suggest that women who have invested a great deal in climbing the career ladder (in science and a variety of other occupations) are choosing to step off the career track, at least for a period of time.[24]

It should not be surprising, in a society where women are the primary caregivers, that many women exit, cut back, or find alternative careers that allow more time with family. Though many of us have no doubt that we have what it takes to succeed in the sciences, children exert a powerful force upon us as well, and we will seek a career that allows for balance. In an article published in *Scientific American* on the effect of pregnancy and motherhood on the female brain, Craig Kinsley and Kelly Lambert write:

> What was once a largely self-directed organism devoted to its own needs and survival becomes one focused on the care and well-being of its offspring. . . . New research indicates that the dramatic hormonal fluctuations that occur during pregnancy, birth and lactation may remodel the female brain, increasing the size of neurons in some regions and producing structural changes in others. . . . Although studies of this phenomenon have so far fo-

21. National Science Foundation, Division of Science Resources Statistics, *Characteristics of Doctoral Scientists and Engineers* (2003), table H-12, http://www.nsf.gov/statistics/nsf06320/tables.htm.

22. Ibid., table H-11.

23. National Academy of Sciences, *Beyond Bias and Barriers*, 3, 36.

24. P. Stone and M. Lovejoy, "Fast-Track Women and the 'Choice' to Stay Home," *Annals American Academy of Political and Social Sciences* 596 (2004): 62–83; Eduardo Porter, "Stretched to the Limit, Women Stall March to Work," *New York Times*, March 2, 2006; also see *First Hidden Brain Drain Summit a Success, The Hidden Brain Drain*, Task Force, Media Notes, Center for Work Life Policy, New York, New York, http://www.worklifepolicy.org/documents/October%202002006%20News%20Flash.pdf; Lisa Belkin, "After Baby, Boss Comes Calling," May 17, 2007, *New York Times*.

cused on rodents, it is likely that human females also gain long-lasting mental benefits from motherhood. Most mammals share similar maternal behaviors, which are probably controlled by the same brain regions in both humans and rats. In fact, some researchers have suggested that the development of maternal behavior was one of the main drivers for the evolution of the mammalian brain.[25]

The changes discussed by Kinsley and Lambert include not only the typical behaviors associated with motherhood—such as nestbuilding, grooming, and offspring protection—but also such changes as improved memory, ability to forage for food, and (not surprisingly for those of us who juggle work, carpooling, doctors' appointments, rehearsal schedules, and grocery shopping) multitasking. Some of these behaviors are long-lasting, benefiting rats, at least, well into their senior-citizen years.

For some women in the sciences and other professions, the pull of family versus career can be overwhelming, as documented by Pamela Stone and Meg Lovejoy, authors of an article entitled "Fast-Track Women and the 'Choice' to Stay Home," who note that "[p]rofessional women are caught in a double bind between the competing models of the ideal worker and the ideal parent."[26] They further observe that "[a]lthough the vast majority of women with professional degrees are working, they are out of the labor force at a rate roughly three times that of their male counterparts and overwhelmingly cite 'family responsibilities' as the reason."[27]

Many of the forty-three women included in Stone's and Lovejoy's study (women formerly employed in professional and managerial jobs) agonized over their decision to leave their jobs, in part because many women are proud of their accomplishments, enjoy their work, and gain a sense of identity through their work.

But seeking balance between career and family shouldn't hobble a career. For example, in the sciences there is a widely held belief that once one leaves the main road, depending on the discipline, the on-ramp can be difficult if not impossible to find.[28] In 1970 Kathleen Lonsdale, an X-ray crystallographer and one of the first women elected to the Royal Society in London, posed the question, "Is it Utopian to suggest that any country that

25. Craig Kinsley and Kelly Lambert, "The Maternal Brain," *Scientific American* 294 (2006): 72–79.
26. Stone and Lovejoy, "Fast-Track Women," 62.
27. Ibid., 63.
28. Xie and Shauman, *Women in Science*, 8.

really wants married women to return to a scientific career when her children no longer need her physical presence should make special arrangements that encourage her to do so?"[29] Thirty-seven years later, women are still seeking a scientific community that will not disadvantage them if they interrupt their careers in favor of family responsibilities but will instead appreciate the breadth of experience that comes with raising children:

> What we all need—parents and non-parents of both sexes—are work places and a scientific community that will accept our bouncing back and forth from periods of work intensity to periods of part-time work. Yes—we as a scientific community will have to be accepting of people needing time to get back to speed. In return we will get mature, balanced people with a wisdom and knowledge of life that would very likely otherwise be missing. (Francesca T. Grifo, PhD, Senior Scientist and Director, Scientific Integrity Program, Union of Concerned Scientists)

Even for those who find balance with a full-time career track, raising a family while maintaining a scientific career is difficult, partly because of biological limitations. What may set science and academics apart from other professions when it comes to having children is the requirement for many PhD graduates to complete at least one postdoctoral position before moving into a more permanent job, particularly for those in research, delaying the timing of career stability. When a woman is striving for tenure or career stability (typically in her early to mid-thirties), the biological clock is winding down. The timing of children and the impact of children on scientific careers (primarily academic) are well documented.

Writing about the conflict between science career and family, Yu Xie and Kimberlee Shauman, authors of *Women in Science: Career Processes and Outcomes*, note that when the primary responsibility for household labor falls on women, some women will forgo their potential science and engineering careers for family, while others who are already on the career track will forgo family for their scientific careers. Xie and Shauman observe that "fewer women than men combine a family life with an active S/E [science and engineering] career."[30]

In his report *Faculty Careers and Flexible Employment*, David W. Leslie presents some striking figures illustrating that as the number of depen-

29. Kathleen Lonsdale, "Women in Science: Reminiscences and Reflections," *Impact of Science on Society* 20 (1979): 45–59.

30. Xie and Shauman, *Women in Science*, 141.

dents increases, the mean number of hours worked by women (presumably work related to academia and not to home and family) and the number of hours dedicated to research decline; by contrast, the hours worked by men tend to increase slightly with increasing numbers of dependents.[31]

Another analysis of tenured and tenure-track faculty at the University of California, Berkeley, by Mary Ann Mason and Marc Goulden in 2002, concluded that women who have "early babies," born within five years of their mothers' attaining their PhDs, tended "not [to] get as far as ladder-rank jobs. They make choices that may force them to leave the academy or put them into the second tier of faculty: the lecturers, adjuncts, and part-time faculty."[32] These findings are supported by the report *Gender Differences among Contingent Faculty* by Elizabeth Ivey and others.[33] In contrast, write Mason and Goulden, those "with late babies and women without children demonstrate about the same rate of achieving tenure, a rate higher than women with early babies. Presumably, women who have babies later in their career life have already achieved job security. They are also more likely to have only one child."[34]

Finally, a survey of approximately one thousand American Fisheries Society members representing a range of work sectors found that "twice as many women as men think having children will adversely affect their careers. For those with dependents (and the study found that women were both less likely to be married and less likely to have dependents), when asked what effects dependents had on their career, more than twice as many women as men said they had put their career 'on hold' because of their dependents."[35]

The experiences of having children and child rearing are different for each one of us. Before I had children, I was clueless about the strength of the mother-child bond. I just assumed that postbaby I'd continue with business as usual while my husband finished his PhD. Two colleagues and I secured funding to investigate the impacts of reproductive contaminants on fish, which meant moving and setting up a laboratory in time for our

31. David Leslie, *Faculty Careers and Flexible Employment*, TIAA-CREF Institute, Policy Brief, 1-06, http://www.tiaa-crefinstitute.org/research/policy/docs/pol010106.pdf.

32. Mary Ann Mason and Marc Goulden, "Do Babies Matter? The Effect of Family Formation on the Lifelong Careers of Academic Men and Women." *Academe* 88 (2002), http://www.aaup.org/pubsres/Academe/2002/nd/feat/maso.htm.

33. Ivey, Weng, and Vahadji, *Gender Differences*, 14–16.

34. Mason and Goulden, "Do Babies Matter?"

35. Nancy Connelly, Tommy L. Brown, and Jill M. Hardiman, "AFS Men and Women Differ Most in Their Lifestyle Choices," *Fisheries* 31 (2006): 503–6.

first field season. For practical reasons, and not necessarily in anticipation of any powerful hormonal pull, I'd written in my salary as half-time for two years. My advisor smiled knowingly when I told him of my plans. "My wife thought she'd stay home after kids, while my sister-in-law knew she wanted to go right back to work," he'd told me as I waddled down the hallway beside him on my last day of work. "Well, my wife went back to work, and my sister-in-law stayed home—so you just don't know how you'll feel." I fell smack between the two, eventually becoming a stay-at-home scientist, determined to have both family and career yet uncertain of my ability to maintain both.

I am not alone in wanting it all. Many women who have devoted a significant portion of their young lives to education and training to become a professional want both family and a career. In her article "The Long Road to the Fast Track: Career and Family," Claudia Goldin explores the evolution of college women's attitudes toward family and work throughout the past century by identifying five cohorts of college women whose experiences reflected the times in which they lived and the groundwork laid by the women college students who preceded them.[36]

The women in the first cohort Goldin identifies graduated from four-year colleges between 1900 and 1919 and are characterized as desiring (or achieving) "family or career"; those graduating between 1920 and 1945 who pursued "job, then family" composed the second cohort; followed by the third cohort, graduates between 1946 and 1965, who tended toward "family, then job." For this post–World War II group, whose members married young and created the boomer generation with their high rates of childbirth, family typically came first, and teaching was the dominant occupation. Goldin writes:

> [This cohort] became the frustrated group described by Betty Friedan. . . .
> [I]ts members became increasingly discontent with a labor market that offered college women little in the way of career advancement and with employment officers who often asked them just one question: "Can you type?"[37]

Although few contributors to this book belong to this demographic, those that do, do not fit the cohort model of "family, then job"—since they pursued a career in addition to raising a family.

36. Goldin, "The Long Road," 25.
37. Ibid.

Women receiving their degrees between the years 1966 and 1979 are the fourth cohort, the "career, then family" generation of college graduates. This is the generation of women who entered a range of professions, some reaching the highest levels attainable. In exchange, at least for some who placed career before family, writes Goldin, "children were put 'on hold,' sometimes forever."[38] Several of the essays in this book were written by women of this generation, but as they describe in their essays, these contributors managed career (primarily in academia) and family simultaneously.

The last generation tracked by Goldin graduated from college in the eighties. Goldin writes that for this generation, "labor force participation when young and married was extremely high—around 80 percent,"[39] and like their immediate predecessors, they entered a diversity of professions. Women in this cohort were cognizant of their biological clocks yet managed, in greater numbers than the previous group, to establish both a family and a career. Women of this generation contributed the largest number of essays to this book and describe a broad range of career paths in government, academia, industry, writing, and teaching and career practices from full-time to part-time and volunteer employment.

Outing the Elephant

Like the cohorts described by Goldin, the contributors to this book, no matter which path they chose to follow, broke new ground and charted new courses in order to achieve an acceptable work-life balance. Some faced few obstacles, while others describe confronting some of the insidious gender bias described by Joan Williams and others in their article "Beyond the 'Chilly Climate': Eliminating Bias against Women and Fathers in Academe."[40] The comprehensive literature review reveals a range of disturbing maternal stereotypes, as reflected by employers who shy away from hiring women in their prime childbearing years, by mentors who suggest that women not discuss or that they hide pregnancies during interviews, and in one study, by participants who favored (by two to one) a childless job applicant over one with children even though their résumés were identical. They report that many who make it past the résumé screenings and

38. Ibid., 26.
39. Ibid.
40. Joan Williams, Tamina Alon, and Stephanie Bornstein, "Beyond the 'Chilly Climate': Eliminating Bias against Women and Fathers in Academe," *Thought and Action* (Fall 2006): 79–95.

interviews and into academia are fearful that taking advantage of family-friendly offerings such as time off for childbearing (or child rearing) might dull one's competitive edge or fulfill stereotypes. This attitude has left administrators concerned that family-friendly policies—such as the part-time faculty positions recently offered at Ohio State University, the University of California, and several other institutions—may fail. These concerns are not unfounded. In a study of chemistry and English faculty from across the country, Robert Drago and Carol Colbeck report that more women than men practiced behaviors that tended to minimize, hide, or neglect family commitments in order to improve work performance or to maintain the appearance of the ideal worker, with varying degrees of success.[41] They write that more women than men reported having fewer children in their pursuit of academic success and were afraid to ask for reduced teaching loads when necessary. The majority of women responding also felt that they'd returned to work too soon after childbirth "in order to be taken seriously as an academic."[42]

Clearly there is room for change, not just in academia but in all sectors of science. In the conclusion of her analysis Claudia Goldin writes:

Each generation built on the successes and frustrations of the previous ones. Each stepped into a society and a labor market with loosened constraints and shifting barriers. The road was not only long, but it has also been winding. Some cohorts of college graduate women gained "family," whereas others gained "career." Only recently has a substantial group been able to grasp both at the same time.[43]

The contributors to this book have volunteered their own stories with the intention of empowering others to speak out not just about their struggles and concerns for the future but also about personal and professional successes achieved while balancing family and a life in science. Contributors write about designing elementary school curricula, working locally as naturalists, and redirecting research toward more applied sciences so that their children's generation might benefit. They write about becoming deans and mentors, of leaking milk during job interviews, maintaining

41. Robert Drago and Carol Colbeck, *The Mapping Project: Exploring the Terrain of U.S. Colleges and Universities for Faculty and Families,* Final Report for The Alfred P. Sloan Foundation and The Pennsylvania State University, http://lser.la.psu.edu/workfam/mappingproject.htm.

42. Ibid.

43. Goldin, "The Long Road," 34.

part-time research, gazing at the earth from afar, and at the universe from earth, writing textbooks, and teaching high school. We view this as a first step in creating a forum in which women who are scientists and mothers feel free to speak about their desires, their goals, and the importance of balancing career and family without being mommy-tracked. In this way, motherhood will no longer be the elephant in the laboratory, and science as a discipline will benefit from the diverse experiences of women who have learned to balance their lives with their work.

SECTION I

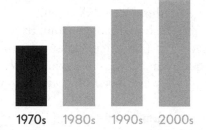

GETTING STARTED

It was winter 1974, and I was tallying the sugars, fats, and proteins that had sustained me for a week. My eighth-grade science class, under the direction of Mrs. Leary, was to keep an *honest* log of everything we consumed—the tuna casseroles, baked potatoes, and meat loaf, along with the M&Ms, coffee milk, ice-cream sandwiches, Oreos, and Ring-Dings (information that I did not share with my mother) over the course of a week. It was in Mrs. Leary's class that I learned an astounding fact: my average daily calorie consumption far exceeded that of my nemesis, Stephen Epstein, the slightly pudgy boy who had challenged me and a girlfriend to a very public two-on-two basketball game à la Billy Jean King and Bobby Riggs. From then on I was hooked on biology.

• • •

By the 1970s, less than a decade had elapsed since the famous newspaper headline announcing that a "housewife" and "mother," Maria Mayer, had won the 1963 Nobel Prize for physics. At some university campuses antinepotism laws still prohibited spouses, primarily women, from holding faculty positions at the same university. Dr. Mayer was a victim of these laws for decades, working without pay at some institutions, because her husband held the faculty position.

A 1971 study published in *Science* entitled "Women in Academia," designed to ferret out bias against women at the time of hiring, revealed "a definite tendency for the [department] chairman to prefer an average male over an average female," although to their credit they were able to differentiate the "superior" female candidate from the average male.[1] The authors were surprised when some department chairmen appended comments to the survey questionnaires, asking what the *husband* of the female applicant would do if she were hired and what *she* would do with her children.

This despite the conclusion of another study, also published in *Science* and based on a survey by the American Society of Microbiologists, that "[w]omen work as hard as men, remain at their jobs as long as men do and basically have the same motivation for working outside the home as men do."[2] However, they also reported that women were more likely to be encouraged to stop their studies rather than pursue PhDs, and that those who persevered nonetheless were discouraged to a much greater degree than were male PhD candidates. Finally the authors reported a disparity in both marriage (44 percent of female compared with 90 percent of male respondents) and family, with 54 percent of women reporting no children while only 12 percent of the men reported having no children.

But the seventies was also a time of change. The Education Amendments of 1972, Title IX, required that educational institutions, from elementary schools to universities, receiving federal funding treat men and women equally. The law applied to everything from sports to college admissions and funding. Equal Opportunity posters depicting male secretaries and female engineers encouraged girls and boys to abandon stereotypes when considering jobs. Congress also passed the Metric Act of 1975, with the goal of converting everything from common household measurements to temperature and highway speeds to the metric system. The American Association of University Professors (AAUP), recognizing the changing times, issued a statement urging institutional changes toward

1. Arie Y. Lewin and Linda Duchan, "Women in Academia," *Science* 173 (1971): 892–95.
2. Eva Ruth Kashket, Mary Louise Robbins, Loretta Leive, and Alice Huang, "Status of Women Microbiologists," *Science* 183(1974): 488–94.

greater flexibility to better accommodate faculty members (particularly women) with children. It stated, "This flexibility requires the availability of such alternatives as longer-term leaves of absence, temporary reductions in workload with no loss of professional status, and retention of full-time affiliation throughout the child-bearing and child-rearing years."[3] Well, temperature is still reported in Fahrenheit; our odometers read miles per hour; liters and grams have not yet transitioned from the laboratory to the kitchen; and the AAUP, along with colleges and universities across the country, is still figuring out how to better accommodate, attract, and re-tain women faculty. Title IX certainly was an advantage for women's sports (and I confess, that prior to researching this book, I thought Title IX *was* all about sports), but what about equality in the sciences, one of the most traditional of "male disciplines?"

Of all PhDs awarded in science and engineering in the United States during the 1970s, only 17 percent were awarded to women, and the fol-lowing essays provide some insight into the experiences of several of these pioneering scientists and mothers. For most of their careers, except for a year or two when their children were young, most of these women worked full-time in either academic or federal research positions. As the reader will see, their experiences contrast with those contributors in later sections, several of whom write about alternative career routes. Perhaps this is a con-sequence of time passing, making it difficult to find those women who chose alternative routes thirty years ago, or, a consequence of the times when fewer options may have been available or even desired by these pio-neers who earned their PhDs in this decade of change.

3. American Association of University Professors, "Statement of Principles on Family Re-sponsibilities and Academic Work," http://www.aaup.org/AAUP/pubsres/policydocs/contents/workfam-stmt.htm.

Balancing Family and Career Demands with 20/20 Hindsight

Aviva Brecher

National Technical Expert on Transportation Safety, Health and Environment, Department of Transportation (DOT), Research and Innovative Technology Administration (RITA), Volpe National Transportation Systems Center

PhD, Applied Physics, University of California, San Diego, 1972

As an over-sixty baby boomer scientist, I welcome this opportunity to look back, share life and career stories, and offer advice to younger women who want to have it all: successful and productive science careers, good jobs, motherhood and a harmonious family life, clean homes, loving husbands, and appreciative bosses. I had a chance to think about which, if any, of my life's "lessons learned" would be of interest or helpful to younger women facing similar career and life choices today.

It is astonishing that they are asking today the same questions about how to successfully manage and blend careers in science with the demands of motherhood and family life that we struggled to solve thirty years ago. While these perennial questions have changed little, if at all, over the three decades since I joined the brave community of female scientists, the personal answers and solutions have evolved. The social reengineering that we dreamed of under affirmative action plans in the seventies has not yet materialized. Women then wanted equal opportunity and equal pay, expecting that carefree parenting and full-fledged tenure-track careers would follow. Yet today we still have to contend with the mommy track and glass ceiling, which limit the pace of our progress in the workplace. Equality at home and at work still seems like a pipe dream.

Study after study published these days finds that the thorny problems that professional women face in balancing family life and career demands are difficult to solve, and that both the academic ivory tower and the business and industry workplaces are still male-dominated, hierarchical, and conservative. Women's work life has improved gradually and cumulatively: we have made progress in achieving professional success and visibility. Now there are more of us in science and engineering. We are also wiser and actually expect our husbands to share the housekeeping workload and parenting responsibilities. Employers are more tolerant and family-friendly, and there are more part-time work options and more day care centers. Above all, there are more of us ready to give a helping hand to newcomers.

Sisterhood is powerful, but it is not enough; it seems just as hard today to plan ahead for family and make career and life choices and adjustments as it was then. It is still difficult today, as it was then, to find good day care and the kind of jobs we need to afford it. Sadly, we have not yet reached parity in status and salary with our male counterparts in the professional scientific world.

So how can we, as women scientists, help one another? Storytelling and sisterhood can go only so far, but an "aha experience," perhaps a new idea, and even a fuzzy, feel-good sentiment can help us realize that we are not alone and that we can prevail as scientists *and* enjoy motherhood.

I find it hard to look back and write about how I successfully combined motherhood, family life, and a science career because in my busy life, things mostly happened without, or in spite of, plans. There was little time to ponder, analyze, reflect, and take stock to gain insight and deliberately change career course. I did so twice, to meet family needs, shifting from academia to technical consulting for higher pay, and later from consulting to government for more family time but at lower pay.

Being a first-generation immigrant from Israel, I may not be typical of American women scientists. When I married an MIT graduate physics student and transferred from the Technion to MIT, there were only a couple of other women in my large physics classes, and most kept quiet. Since I was used to a brash culture, I dared to ask questions and speak up in class and even requested help from some professors, who kindly took me under their wings as mentors. This bold attitude and directness helped me compete with male colleagues throughout my career.

As I reached my thirties, the biological clock's ticking accelerated, along with parental demands for grandchildren. My husband and I remained unconvinced that it was advisable to have children. Two other physicists provided us with role models, which helped sway us. My mentor and role model at MIT (Institute Professor Mildred Dresselhaus) was married to a physicist, with whom she collaborated on research and shared parenting duties at home. Their four children were accomplished musicians and wonderfully well behaved, and they shared in household duties as well. Over a dinner at her house, Millie told us stories about how she brought her young children to the lab, where they played while she conducted experiments. The following advice offered by this successful professional couple resonated with us: "Children must come in pairs for harmony in the household, just as paired electrons with spin up and spin down can form a stable atomic orbital. One or three children would not work as well as two or four!" I was inspired to follow her lead but cut it in half to match our energy level and budget limitations.

Another high-power couple, an MIT Nobel laureate in physics and his Harvard lawyer wife, also offered a convincing rationale for becoming parents: "When the whole world hates you at work, or ignores you, a child will enrich your life with unconditional love." (Several years later we figured out how soon this love becomes conditional!) Within two years of these conversations—after completing our PhDs, and when we had relatively stable jobs and could afford a house with a yard, as well as babysitters—we had two children, a boy and a girl, two cars, and a nonstop roller-coaster home life.

In our case, the arrival of children was equivalent to an earthquake or tsunami: it drastically changed our outlook on life, our rhythms, and our interests. Before becoming parents we worked on esoteric astronomy topics, my husband on the universe (cosmology, quasars) and I on the solar system scale, on meteorites, and the Apollo lunar missions. In recognition of this earlier work, the International Astronomical Union named Asteroid 4242-Brecher, which is located between Mars and Jupiter.

After the children arrived, we both became more grounded in our interests: my husband began improving science and math education, and I worked on real-life problems like nuclear waste disposal, climate change, and advanced transportation systems safety to leave a legacy of a cleaner environment and to develop better transportation for a safer world.

As a young postdoctoral research scientist at MIT, I had so many career pressures that time out for motherhood was not an option. After two weeks at home, our daughter spent her first three months in a bassinet on my desk at MIT, attending seminars while breast-feeding under my poncho. After ng "graduation" she started going to day care nearby at the home of an MIT graduate student, whose wife was a new mother like me. Similarly, three weeks after giving birth to our son, I started a new job as assistant professor of physics at Wellesley, teaching two courses and two labs each semester for the next three years, while continuing my research at MIT. We then hired an experienced mother to come to our house and babysit for our toddler and newborn so we could continue our faculty careers.

My work as a research scientist in fast-moving fields also did not allow me to take time out to raise our children in their early years. Additionally, we could not have afforded to have two children on a single academic salary. I once tried consulting out of the house for a year after our son complained that I was not a "normal mother" who would be at home when he got back from school. That year I was still in pajamas working to complete assignments when the kids came home, having worked nonstop since they left. The money too was tighter, in spite of saving babysitting fees, and the workload was unpredictable, so I went back to full-time work and two reliable incomes.

To my mind, the key to children's happiness and parents' peace of mind is to strive for stability in career and home life and to compromise, compromise, and further compromise with spouse, babysitter, boss, and children. Fortunately, my husband has always shared parental duties with wonderful dedication and enthusiasm while maintaining a full and uninterrupted academic career with lots of travel. We also lived in the same house for over thirty years, a real family home.

When we moved the whole family to Washington for a year so I could serve as American Physical Society (APS) Congressional Science Fellow, my husband was able to secure a National Research Council Fellowship at NASA, and the children thrived in this new environment.

I believe that the most basic requirement for a successful career and family life is to have a good babysitter whom you not only pay well but also value and respect. Parents must secure reliable backup child care in case the babysitter gets sick or you have to travel overnight and over weekends, or even for months at a time (as we did, I for a research fellowship in Japan and my husband on an exchange visit to China). Since my family was in Israel and my husband's in New York, while we lived in Boston, we could not get help from our extended families. We were fortunate to find and keep for thirteen years a wonderful "second mother" who took care of our brood during the working hours and whose older children could double as overnight sitters in times of need.

I suspect that each mother has her own secret for balancing children's demands with other home and work-life pressures. My secret was bringing home presents from my work trips to please the kids and appease my guilt from not always being there. So I ended up buying them too many gifts or being too permissive to make up for it, which they fully exploited. We developed by default a good cop–bad cop approach to child discipline and psychology. I was usually strong on theory, always reading books about parenting, but weak on practice. To balance, my husband was more level-headed, using common sense and a "fly by the seat of your pants" mentality. I was less patient and more willing to bribe the children if needed to modify behavior, while he was big on persuasion and reasoning. Needless to say, our children could and did catch on pretty soon and started to play the two ends against the middle. Unfortunately, by the time we acquired good parenting skills, the kids were out of the house and driving our cars, college-bound.

I can now proudly say that we have successfully raised two well-adjusted, responsible and accomplished young adults. We have somehow survived the strains of time and stages of growth—from the terrible twos to terrible teens, from college admissions to an empty nest and dual tuition payments. We have been able to provide them with a stable family environment in a nice middle-class home and with a good education. Perhaps we made a mistake in overexposing them to science: science talk, science books and encyclopedias, lots of science museums and science fair projects, hobbies like rockhounding and fossil collections, to the extent that despite having taken plenty of math and science courses in high school and college, both our son and our daughter have opted for nonscience careers. Perhaps they saw how hard we worked for rather modest salaries and decided to do better?

Over our dual-career life, as we navigated unavoidable professional transitions, job changes, moves, and new interests, our children served as our center of gravity and as a touchstone for what's real. Though the cost in labor, time, fatigue, aggravation, and real money is scary, there are great compensations and returns on investment like love, warmth, and respect from our children, with even greater future dividends hoped for as we get older.

With 20/20 hindsight, the life metaphor that comes close to describing how we managed to combine career and family is that of a peaceful blue pond in a summer breeze before the children arrived, followed by stormy seas and turbulent weather after their arrival. Or maybe a better analogy is running on a balance beam, holding four shopping bags in your hands and trying to check your watch and hurry home to make dinner. The feeling that I recall from my early years as a working mother is a heady and perennial mixture of bone-deep fatigue and steely discipline and determination, leavened with exhilaration and wonder, joy and anger. My husband and I have been married now for more than forty years and have survived as a dual-career "pair of docs"—but only with unavoidable career and lifestyle compromises.

I firmly believe that each of us must define professional success in personal context, according to our needs and circumstances. In retrospect, I certainly did not move up the professional pyramid as fast or as high as some of my mentors and MIT classmates, perhaps because I had more modest or more realistic aspirations, balanced with personal satisfaction from achieving a "normal" family life. I have, however, succeeded in maintaining a productive, interesting, and very satisfying career and family life over several decades, in spite of dual-career constraints and choices favoring family over professional advancements. I have always been willing to serve in professional societies and on numerous committees over the years to contribute to good causes, and I was recognized for dedicated service as a Fellow (by the American Physical Society [APS] and the American Association for the Advancement of Science [AAAS]). I felt part of a community and was able to mentor younger colleagues. I am quite satisfied with my family and professional life accomplishments, even if I did not do Nobel Prize-winning, "important" work.

So, yes, it can be done! You can combine and balance family stability and career demands, but at a steep price. You need good planning to secure a triply redundant emergency support system for child care, negotiating skills for good spousal support and an ideal babysitting arrangement, and acceptance of sacrifices in both personal time and professional expectations. It takes willpower, discipline, and persistence. Do you have the willingness to sacrifice? The gumption to accept disappointments? The thick skin to not take insults at work personally and persevere? The courage to face the reality that you cannot be a supermom and a superscientist at the same time because, temporarily, something's got to give? You must have persistence, foresight, and energy to live through long days and sleepless nights and the willingness to give up personal time and put family first. Personally, I believe the rewards are worth the high price, but the agonizing and not always rational choice must be yours alone. Good luck to you!

Extreme Motherhood

You Can't Get There from Here

Joan S. Baizer

Associate Professor in Physiology and Biophysics, University at Buffalo,
State University of New York

PhD, Psychology, Brown University, 1973

I seem to have lived my life out of order. I married, divorced, and had a baby, in that order. And now, at an age when many colleagues are contemplating retirement, I am as engrossed in my work as a postdoc.

I was married at twenty-one, three days after graduating from college in 1968, to a high school boyfriend. I had become very interested in neuroscience and knew I wanted to do a graduate degree, but my choice of graduate school was determined by his choice, not atypical for 1968. We stayed together through graduate school and postdocs but separated in 1976. I was thirty and on my own for the first time. I took a job as assistant professor in a medical school department. I very much wanted children and assumed that husband number 2 would arrive promptly. He did not, and after much agonizing I decided to have a child on my own, through carefully planned donor insemination. I did not discuss this decision with the department chair or other officials; I just started showing up at work pregnant. I think many people thought that the father was a friend and coworker—he and I were often seen together—but the father was in fact IDANT #524. I had all the prenatal testing appropriate for a forty-year-old first-time mom, with no sign of difficulty. Jessica was born in 1987.

I had already gotten tenure by the time she was born (just barely) but needed rather badly to rejuvenate my work and wanted to make some changes in my research direction. To that end I arranged a sabbatical in 1988 at the National Institutes of Health. Since the people with whom I chose to work also had a small child, I was confident they would be sympathetic to the constraints imposed by child care. During that year I found out that Jessica was "developmentally delayed," a euphemism for not meeting developmental milestones like sitting, crawling, walking, and talking. But I was able to make good use of the sabbatical, finding family day care for her and working a normal workweek. I returned home to Buffalo hoping to continue research. However, I had major concerns about Jessica and continued to try to find a diagnosis and prognosis for her, as well as an appropriate preschool and therapy. She was not diagnosed until she was two, after many different doctor visits. The diagnosis was a rare disorder called Hypomelanosis of Ito, which causes abnormal brain development. In her, this means mental retardation (at nineteen she is cognitively about age four), epilepsy, and "challenging behaviors," the politically correct disability term for saying NO, hitting, kicking, scratching, and other unfriendly behaviors when frustrated or angry. As a very young child she was probably no more difficult than a typical child, and my life was constrained only by when the day care center opened and when it closed. In addition, I needed to take her to doctor's appointments, physical therapy, and speech therapy sessions. As she got older, however, she began having frequent seizures not controllable by drugs and exhibited much more difficult behavior. A fighting two-year-old in a teenager's body is very tough. Arranging appropriate school and after-school care became extremely difficult, and negotiating these arrangements took me out of the laboratory countless times. Seizures and "behaviors" are the most difficult of the disabilities to manage. Caregivers are terrified of seizures, and no one wants to be hit or kicked. I was often called to pick her up from school or an after-school program because she had just had a seizure and could not go on the bus. There were several calls to meet the ambulance at the emergency room because she had been injured in a seizure fall and was bleeding and needed stitches. Her seizure disorder often resulted in interrupted sleep for both of us. Seizures were not the only problem: sometimes I would be called to pick her up because she had hit someone.

During this time I continued with my full teaching load. I never once canceled or was even late to a lecture, although that often meant extremely complex (and expensive) arrangements for care. It never occurred to me to ask for a reduction in teaching. I felt that fulfilling my teaching assignments was my responsibility and that personal circumstances were not relevant. It was a revelation last year on a committee when I read a promotion letter from a chair explaining that a candidate's teaching load had been reduced because he was the single parent of an autistic child.

Obviously Jessica's care severely limited research time and energy, and I have remained an associate professor for more than twenty years. Her care also limited my ability to travel to meetings or to visit other universities, since overnight care for her was practically impossible to find. I often said that if I had had any other job, I would have been fired! The academic schedule at least let me keep working. I finally found an adequate group home placement for her when she was seventeen (only, I should say, through the intervention of my New York State assemblyman). I did this with great reluctance, since one is supposed to care for one's child, no matter how difficult, but I was simply unable to function any longer. Since she has been there, I have renewed my interest and involvement in research, and I am now working in a way that most people do early in their careers. I still have her home one night a weekend and take her out during the week, but I can come to work without feeling as if I have already fought World War IV before leaving my house, and I am able to stay late many nights.

I should also say that my experiences with Jessica have influenced my life and my career in positive ways. First and foremost, taking care of Jessica and meeting other parents of disabled children has given me a perspective on what is truly important. A few years ago I spent a morning as a parent advocate on a Committee for Special Education meeting. One case that was discussed was that of a boy who was not doing well in school and seemed depressed. Did he need a special education class? He was living with his grandmother. His mother was dead and his father in jail. In fact, his father was in jail because he had killed the mother. This boy clearly needed a whole new life, not just special ed. In a dramatic contrast, the next night I had dinner with a friend who is in the math department, and she told me all about a dinner with a visiting speaker, at Buffalo's best restaurant, and her great distress over the fact that the chairman, rather than the speaker, had ordered the wine. The contrast in perspective was stunning.

Similarly, I think my view of what is valuable in people has been radically changed. I come from a Jewish intellectual background in which one's worth as a person is considered roughly equivalent to one's IQ. I selected the sperm donor to be "bright" and really expected to have a "bright" child. Knowing Jessica and her peers has made me see many more dimensions to human value than IQ.

All the advocacy I have done for my daughter greatly increased my social and negotiating skills, in addition to enhancing my ability to engage in scientific collaborations. I can also cope in the world much better than I could before. I have learned to be persistent and firm while remaining pleasant and nonconfrontational, and that skill has been as useful at airports and hotels as it was at school meetings. I now never, ever leave anything (grant applications, lecture writing) to the last minute, since for so many years I could not count on there being that last minute—maybe I would be called to the emergency room. Setting an arbitrary deadline for myself that is days or weeks before the real one is actually a very effective way to work.

I have been extensively involved in community service advocating for the disabled and especially in parent advocacy. I have met and worked with a spectrum of people with whom I otherwise would have had no contact and have learned how to translate what the "experts" say to the level of a loving and concerned but uneducated parent. My teaching has been informed by my experiences with disabled children. Lectures to the medical students include discussion of cerebral palsy; lectures in a pathophysiology course include coverage of epilepsy and some other developmental disabilities. Conversely, my teaching experience was very valuable in the advocacy work. I have several collaborative research projects that resulted in part from my curiosity about brain development and what can go wrong with it.

Just the other day I went to a farewell reception here for a young faculty member who was leaving for another job. The department chairman gave a little speech in which he characterized the man as "the ideal faculty member" because he always worked until late at night, every day, seven days a week. Under this definition neither I nor mothers of much easier children nor men who want to be involved fathers would be considered "ideal." I hope the standard changes!

Careers versus Child Care in Academia

Deborah Ross

Professor, Department of Biology, Indiana University-Purdue University

PhD, Microbiology, Rutgers University, 1974

I decided that I wanted to be a biologist in junior high school, following my first exposure to a real biology course. My parents had always encouraged me to excel in science and mathematics. My father was an electrical engineer who tended to have science-based hobbies such as rock collecting and astronomy. My mother had a master's degree in medieval English history, unusual for a woman in the 1920s. After I graduated from grade school, she became a substitute teacher for English and history classes. Although she enjoyed more traditional pursuits like sewing and embroidery, she was an avid bird-watcher and rock collector, one of the few interests she shared with my father. As a family, we would go on trips to look for condors in the Ojai Valley and geodes in the Mohave Desert. Under my father's supervision, I ground a mirror for a four-inch reflector telescope and made an AM-FM Heathkit receiver. Because of this background, I was predisposed to take the advice of an inspiring biology teacher who told me that microbiology was an exciting field. My parents bought me a microscope and I was off straightaway to hunt microbes!

I never questioned whether I was capable of doing science or whether I ultimately would be able to make it a career, even though my peer group thought that careers for women were a waste of time. I remember quite vividly a discussion in my senior economics class (in 1963) on the subject of whether it was worthwhile for a girl to pursue a college education. The conclusion was that it was worthwhile because the woman would be able to get a better, higher-income-earning husband and to help her children with their homework. I thought this was nonsense!

I continued my education and received a BS. I never met with real opposition until I started work on my master's degree at Cornell University. I was the only female graduate student in the agronomy department at the time (I was not allowed to be a member of the agronomy club), and my advisor made it clear that I couldn't expect a real career in science because of course I would eventually get married and have children and would devote my time to my family. He had never had a female student complete a PhD. Naturally I ignored him!

After finishing my master's degree, I transferred to Rutgers University, where I was one of two female graduate students in the department. The chair of the department informed me that there had been only one female graduate student who had completed her degree since the 1920s. (She presumably gave up her career to marry Robert Starkey, who became chair of the department in the 1950s.) One other female graduate student had quit after two years to get married. Again, I let this information slide off my back and went ahead on my research project, all the while ignoring the nude pinup in the 37 °C walk-in incubator. I eventually did become the first female in forty years to receive a PhD in the department and went back to Cornell to do a postdoc. I found that some things had changed: there were several female graduate students in agronomy, and they were now allowed to become members of the agronomy club.

Up until this time, I had assumed that I would have a career, although I didn't really investigate what my options were. During the second year of my postdoc, I started applying to both industrial and academic jobs. None of the academic positions offered me so much as an interview, so I was delighted when, after a visit with an industrial recruiter and invitation to visit Cincinnati, I received a job offer with Procter & Gamble. I accepted happily and went to Cincinnati, where I eventually met and married my husband. At the time, P&G had some of the most liberal child care policies (although other companies have now caught up). I had a full six months' maternity and child care leave, for which I was very grateful because it really gave me the opportunity to devote my time to my infant daughter. When I went back to work, my husband was temporarily between jobs, so he took over child care for a few months. By the time he went back to work, she was old enough to place in day care.

As far as raising a family was concerned, my position in industry was ideal. I worked a forty-hour week, with liberal policies regarding sickness and medical appointments. We were not expected to work late or on weekends (in fact we were actively discouraged from doing so), but of course we could take work home with us, like report writing and so forth. Thus my job very rarely interfered with my family life. I was becoming more and more discouraged with the work environment at P&G, however, and especially the projects to which I was assigned, because they were getting farther and farther afield from my expertise. I eventually decided to look for an academic opening.

I was hampered in my new job search by a lack of publications (which is typical for those who have worked in industry) and the fact that I could not report on much of my research because it was confidential. I had to go back to my PhD and postdoc work when giving the research seminar required as part of the job interview. At Indiana University-Purdue University Fort Wayne, however, since I was applying for a position in environmental toxicology, my time in industry was considered an advantage. I was offered the position and was happy to accept it.

I was totally unprepared for the amount of time it would take to teach new courses and to develop a research program. I found that two days a week I was working ten hours at the university; it was the policy of the department chair to schedule dual undergraduate-graduate courses, such as my environmental toxicology course, in the evening, either at the 4:30 or 6:00 time slots. Even on these heavy days, I would bring another three to four hours' worth of work home. In addition, I worked all day Saturday in the lab and most of Sunday at home grading papers and preparing lectures.

Naturally this put a crimp on my family life. My husband initially had difficulty finding a job in our new city, so he was able to take over as primary caregiver, but this also added a strain because his role as a stay-at-home dad was very unusual in Fort Wayne at the time (it still is). However, my husband was aware of how important it was for me to work toward tenure during the first five years of my appointment (we go up for tenure during the sixth year). I had a lot to make up because I had no research project that I could bring with me, so I had to start afresh. Right away this put me at a disadvantage by comparison with other new faculty who could continue working on a postdoc or PhD project. Moreover, there is usually a period of time in the tenure track when publications are still coming out from earlier research. Even though this research is not conducted at the university, it still looks good on one's resume to have had a continuous stream of publications rather than a large hiatus.

This workload continued into the second semester of my first year, when I had a new undergraduate-graduate-level course to develop and two biology labs for nonmajors. In my second year, I was told by the chair that the business school wanted the biology course taught at 6:00 instead of 4:30 so their part-time student could take it. I didn't think I could refuse, nor did I think I could ask him whether both sections of the course had to be taught by the same instructor. I think at this time I was still operating under the industrial model, regarding the department chair as "boss." I could kick myself now for not at least raising the possibility of splitting the load with another instructor and pointing out that I was the only professor in the department who was teaching four evenings a week. This schedule meant that I could not attend any evening school functions for my daughter, including parents' night. I also couldn't participate in any of the parents' activities during the day. At this time, my daughter was attending a private school. All the mothers would get together on a regular basis to plan support activities. Very few of these women worked, and they regarded someone who did as a form of subhuman—I did ask once why they didn't have some of these sessions in the evening and was told that that would cut into family time!

As the tenure decision approached, I was also beginning to be very worried about whether I would measure up in terms of publications. In a yearly progress meeting with my chair, he blithely mentioned that I should have six publications by the time I went up for tenure (one for each year, including the sixth, although we go up for tenure at the beginning of the sixth year). I pointed out to him that this would be impossible for me, so I might as well give up now. He backed down and agreed that if I had four, with evidence that others were in the pipeline (under preparation, submitted, or in progress as evidenced by abstracts at national meetings), he would support my tenure case. So I continued my research program, accumulated publications, and eventually received tenure. Right after the tenure decision, the chair left to accept a position elsewhere, and the new chair had his own agenda for the department, but the faculty's teaching load was not one of them.

By this time, however, the long hours had taken a toll on my marriage. My husband and I went through a period of separations followed by brief interludes of reconciliation, but after two years of seesawing back and forth, we divorced. Although we had joint custody of our daughter, she was living with me and I assumed responsibility for picking her up after school and taking her to events like Girl Scout meetings. This meant I simply couldn't teach in the late afternoon or evening. Fortunately, the chair was accommodating. Although this did cut into my research productivity, I received grants that allowed me to hire undergraduates to work in my laboratory. I felt that I was making progress, and I believed that my annual raises, although modest, were at the high end of what was being given to other, male professors in the department.

A new problem emerged, however: what to do with my daughter during summer vacations, especially since I tried if at all possible to get summer grants that gave me support as a guest in a colleague's laboratory. To solve the problem, I relied on creative juggling between my temporary place of employment, Fort Wayne, and summer camps.

Several years after receiving tenure, I began to think about promotion to full professor. Upon consultation with the chair, I was told that if I obtained two additional publications, then he would support me for promotion. Armed with this promise and with a sabbatical coming up, I made plans to spend several months out of town. By now my daughter was away at college, so availability of child care was no longer an issue. My husband (my ex-husband and I had remarried) was able to take a leave of absence from his job and join me for part of the time. I was able to complete the research for two papers.

A year after my sabbatical, I felt confident in putting my case forward. I was surprised, however, when I learned that two other professors, both junior to me and with fewer publications, were also preparing cases for promotion. It seemed that the chair had different ideas of what constituted a case for full professorship depending on the candidate's gender. As the cases traveled up the academic hierarchy, it also became clear that the administration put more value on grants than on publications.

If I sound bitter, it's because I am. I think I have given up a great deal over the years but have not received adequate recognition for it. I enjoy teaching, and my students think I am good at it, but this is not sufficient to receive promotions or above-average salary increments at my institution. My research isn't of sufficient quantity to receive adequate recognition from my peers in the scientific community.

But I'm not so sure I would have done things differently. What about the lost years of my daughter's childhood? Maybe giving her the attention that I could, instead of all of it, worked out just fine. I'm not aware of compelling research that demonstrates that time spent in day care detracts from a child's well-being. My daughter and I now have a very warm and caring relationship. She is independent and self-confident enough to go off on her own to teach English in France and Japan. I still wish, however, that I hadn't had to choose between my daughter and my career during my years before tenure.

Identities

Looking Back over Forty Years as a
Social Scientist, Woman, and Mother

Marilyn Wilkey Merritt

Associate Research Professor of Anthropology, George Washington University;
Lecturer in Anthropology, Catholic University of America

PhD, Linguistics, University of Pennsylvania, 1976

A scientist of the so-called soft variety, I write as an ethnographic linguist—
after forty years of motherhood, a decade of being a grandmother ("Mere-
mom"), and twelve postdoctoral years living overseas. I grew up in America's heartland, in small town midwest Missouri, where my encouraging
family and teachers made anything seem possible. I moved away for higher
education and early on embraced the identities of wife and mother, only
gradually sensing the situational threats to my emerging professional
identity. By my thirties the feminist mantras of "Be the best" (impossible
without opportunity) and "Have a supportive husband" (even the most
supportive husband has goals of his own) were proving unworkable and
wearing thin.

Managing logistics and emotional support for a spouse and children is
something most of us learn on the job, with few useful models or guide-
lines. Planning is good but not consistently realistic as life happens. Flexibility and creative solutions are essential.

At work, luck and informal know-how often help, and sometimes as-
sertiveness or unfettered naiveté can open the door of opportunity. Success takes persistence and enormous psychological energy. The confidence
to keep striving requires the acknowledgment of life goals and genuine appreciation of all that been accomplished.

Too often professional advancement leaves a residue of personal sacrifice or compromise. As I somewhat enviously applaud a group of old friends from graduate school, all full professors at major universities, I note that not one of them is both a mother and still married to the same man. The accomplishments of my own itinerant eclectic career are modest but meshed with a happy marriage and having two wonderful children.

Gary and I found each other early, while in college, during a summer when we both had jobs in magnificent Colorado. After long months of separation (I was in school in Chicago, he in St. Louis), handwritten letters, a single visit each semester, and Christmas engagement, we were married Labor Day weekend following our junior year. I relinquished my Northwestern University scholarship, negotiated senior standing at Washington University in St. Louis, and secured a student loan and part-time job so we could be together.

After a triumphant graduation in May 1963, we searched for graduate programs and jobs and finally settled on a program for Gary at Washington University and a civil service job in St. Louis for me to pay expenses (these were the early days of the Vietnam War when Gary was subject to the draft). Determined to keep some momentum in my own intellectual career, I gained admission to the graduate program in anthropology and took one course each semester. A year later Gary urged me to pursue my studies full-time. Fortuitously, our apparent pluck resulted in two modestly paid research assistantships. But by the time I was finishing my master's in anthropology, I knew I wanted to pursue linguistics as more than a subfield of anthropology. Since we needed to stay longer in St. Louis for Gary's studies, I blithely reasoned that this was a perfect time for a baby.

Wow! Having married at twenty, I was becoming a mother at twenty-four. Luckily my professors still believed in my abilities and hired me to teach my first university course while pregnant, continuing as a teaching assistant the following semester. Fortunately, my professional duties required only a few hours a week away from home and our beautiful daughter, Brienne. Then, when the baby was three months old, we moved to Philadelphia, where Gary took a tenure-track position in sociology (while also writing his dissertation), and I began my doctoral program in linguistics at the University of Pennsylvania. With the baby and no financial support for my studies, I enrolled in one course that spring. But with higher expenses in Philadelphia than St. Louis, I also taught anthropology at night. I found another graduate student mother with whom to share daytime babysitting, and my studies went well enough that I was awarded a

National Science Foundation traineeship. Linguistics at Penn was an intellectually intense program with well-established professors who didn't always agree and who were all men. Nevertheless, they all prided themselves in being socially progressive and genuinely encouraged women students.

The seven and a half years in Philadelphia were tough but also rewarding and formative. We bought a hundred-year-old row house that needed extensive remodeling. We embraced urban renewal and often hosted several wide-eyed neighborhood children. I studied at home when possible, but it was hard with the distraction of children and do-it-ourselves construction. It took three years for me to finish course work, pass doctoral exams, and select a dissertation topic. Then we decided to have a second baby—our delightful son, Seth.

By that point my NSF traineeship had ended, so I supplemented our income with part-time teaching again (two of my former students from those days are now professors of linguistics at prestigious universities). Most of my time, however, was spent pursuing the lonely work of research and writing on discourse analysis in everyday social interaction. Even my two-year-old learned the phrase "working on her dissertation." The whole family went to Michigan for a summer institute in sociolinguistics, where, for the first time, mothers could schedule both course work and cooperative group babysitting.

Then Gary was invited to join the United States Agency for International Development (USAID) in Washington, DC. With me clutching a draft of my dissertation, we accepted. Exciting applied work for Gary and a less stressful neighborhood made the choice easy for us.

We arrived in Arlington, Virginia, in our early thirties, with two children and one good government job. To most of our new friends, I was a suburban housewife with small children (Brienne was seven, Seth only three); indeed, I relished children's outings and informal discussions with other women. I also set a rigorous schedule of morning work. I contacted other linguists and anthropologists and became active in professional clubs.

After I was awarded my PhD, things began to come together. With both children in school, I published a major article and attended conferences, wrote an educational research proposal at the Center for Applied Linguistics, and took a one-year replacement position at the University of Maryland. My approved proposal launched me as principal investigator—studying language from videotapes of primary school classrooms with children aged four to nine. My experiences and intuitions as a mother were invaluable assets.

By 1980, however, the project was winding down, and I was not selected for a rare local academic position. More soft-money research was the most viable professional pursuit, and I naively speculated this could be done anywhere with equal ease. When Gary was offered a foreign service posting in New Delhi, India, the exotic allure was irresistible. We studied Hindi, rented out our house, packed our bags, and said good-byes. Our children would finish eighth and third grades before we left. I completed my project with good prospects for book publication and gave papers at conferences. I was startled when a researcher friend embraced me with "Have a nice life!" but his words would echo through the years ahead.

India was fabulous. Besides learning Hindi, I threw myself into cultural events and traveled extensively (including professional trips to other parts of India). I joined the India International Centre, became an officer of the International Women's Club, served on the school board, wrote a magazine article for the U.S. cultural affairs agency, and consulted on the embassy's Hindi language program. I met Indian linguists, anthropologists, journalists, feminist activists, writers, and artists. In three years I spent more on my activities than I grossed, even with three trips funded by conference sponsors. In between I kept writing on my little typewriter. The children were busy and happy; life as a mother and wife was good.

I befriended other American and expatriate women who felt their personal (professional) identities languishing from not having paying jobs (except for a few locally hired school teachers) and always being seen as someone's mother or someone's wife. Brought up in a money culture, many felt a declining sense of self-worth even as they enjoyed privileged lives. Those who had not completed their education or who came from a country other than the United States felt particular challenges. I knew Indian women with more serious issues. (It was the era of "dowry burnings," when, ironically, the outlawing of dowry resulted in unregulated greed and the "accidental" death of many young brides who were replaced when their families could pay no more.) We all wanted to support one another, but we had different needs, desires, and expectations. These alignments were now part of my identity.

After three years in India, we were assigned a four-year posting in Nairobi, Kenya. We had gone to India as a family and were going to Kenya as a family again, but this time we knew our first year there would be Brienne's last year of high school, with college looming half a world away. Washington University in St. Louis was close enough to my family for her to visit them on long weekends, and I insisted on accompanying her back (paid for with a rare editing job), but leaving her there was truly heartrending. Everyone helped me through it. Two friends with older sons regaled me with how crazed they had been by the "college rupture" and their newest career involvements.

In 2007 it's hard to appreciate how much improvements in communication technology have changed the experience of being at such a distance. Seeing our daughter only twice a year was incredibly difficult. Letters took two weeks each way, telephones were unreliable (roommates shared phones), and there were no cell phones, fax, or e-mail. I would also learn in untimely and difficult ways about the death of my dissertation mentor, the demise of my book on classroom research (my publisher suddenly went under, and a second took a year to decline), and other long-distance publication snafus.

I was often discouraged, but I kept putting myself out there. Over the four Kenyan years I continued writing (including an article on the American ambassador's wife and her personal accomplishments); pursued women's empowerment activities (the 1985 Decade for Women conference was in Nairobi); gave presentations on cross-cultural communication; served on the board of the international school; became a guide at the National Museum of Kenya; taught at U.S. International University; consulted as a social science analyst for USAID; developed African colleagues at the University of Nairobi with whom I coauthored papers, conference presentations, and research prospects; and collaborated with Canadian and African colleagues to research multilingualism and science education in Kenyan primary schools. We traveled throughout East Africa's spectacular landscape (lots of camping) and made lasting international friendships.

The plight of expatriate women (and some men) struggling for professional identities and personal accomplishments as the accompanying spouse was of continuing concern. I helped organize a support network for sharing interests, contacts, and opportunities, paid or unpaid. Most were starving for chances to contribute knowledge, practice skills, learn new things, and simply participate in their passions.

Wherever we are, we feel professionally stymied when there are fewer opportunities to participate, lacking a relevant job or unable to attend conferences. Even when unpaid participation can be negotiated, it takes enormous psychological energy to always be the initiator risking rejection. For me it was important to keep listening, to be part of the dialogue, and to nourish my passion with substance and the physicality of "being there."

Then, in 1987, after seven years abroad, we returned to the United

States and our old neighborhood. The work and reverse culture shock were daunting and exhausting. But we were happy to be home for Brienne's last year at the university and Seth's last two years of high school, to see more of our families, and to renew life friendships. Back in Washington, there were many good universities, museums, and institutions with intellectually stimulating public events. I worked on the Kenyan data, gave presentations, and looked for openings.

There were six uncertain years before Gary was posted again overseas. In my second year, I taught as a visiting faculty member at Georgetown University. I was interviewed at another university for a tenure-track position, but my heart sank when someone asked, "But doesn't your husband travel a lot?"

Professional losses were amplified by Seth's departure for a two-year program at Deep Springs College in the California desert, some three thousand miles away. Fortunately, Brienne returned to Washington, living nearby with friends. Two years later, my father died. Our lives as a family were in transition.

Wounded by academics, I turned to broader interests in writing and creativity and wrote commentary for a sculptor and long-time friend from Philadelphia days. I wrote more poetry and spent more time with art and nature. Networking and meeting old friends from overseas helped me to value experiences at foreign posts.

Eventually I learned of the diplomacy fellowships offered by the American Association for the Advancement of Science. I wrote the strongest application I could muster and returned to part-time university teaching. When chosen to work as a AAAS fellow in USAID's Office of Education, I was ecstatic about the prospect of starting and learning something new and of spending two years (1990–92) applying my academic training and cross-cultural living experience to international policies during the United Nations Decade for Basic Education.

My personal experiences were reinforcing what I had learned as a linguist and social scientist. As our lives unfold, we all have complex multiple identities reflecting changing roles; these are instilled through social interaction and typically involve particular repertoires and uses of language. When we use the vocabulary and argumentation of our professions, we invoke those identities, reinforce our sense of belonging, and nourish that aspect of who we are (or were). In professional contexts, we feel affirmed when others acknowledge our hard-won credentials and slighted when a competing identity is invoked instead (e.g., Mrs. rather than Dr.). Without conversational occasions to appropriately enact our scientist selves, there is an almost inevitable erosion of confidence in that identity. Abroad—where boundaries often depend on nationality and organizational position rather than qualifications and desire—I observed my own disempowerment and that of other women who were denied participation in their fields (including lawyers and doctors). It was strengthening to at least talk among ourselves.

By early 1993, just as I was beginning a semester of part-time teaching at American University, Gary had a confirmed two-year assignment to Niamey, Niger, leaving in October. Since we would be going there as a couple without children (and the many contacts and activities they provide), I was determined to make this a professional assignment for me as well. I applied for a Fulbright opening (not awarded that year), reached out to UNICEF and USAID contacts, and studied French.

Our children made the most of the remaining time we had in the United States. Brienne was now engaged, and we prepared for her July wedding; she and Rusty would live in the Washington area. Seth came home for a summer internship; he was now an undergraduate at Washington University in St. Louis but no longer on our travel orders.

Niamey is a remote and expensive destination with a harsh, baking climate. Our tour stretched from two to three years, with Seth visiting only once and Brienne not at all. The separation, with limited communication, was a real strain. I had no access to e-mail and neither did Brienne (just before wide-spread residential e-mail and Internet services); Seth had access only through the university computer center; Gary's USAID e-mail was restricted to government use except for emergencies. Fax didn't work for either of our children. There was a seven-hour time difference, and telephones (there were no cell phones) were outrageously expensive. Even with official "pouch" mail, letters took two weeks each way. We loved the Nigerien people and the international community, the beauty of the austere landscape, and the importance of the development work, but Niamey was a true hardship post.

By some kind of mother-daughter telepathy I telephoned Brienne the day she found out she was pregnant. I returned to the United States for R & R in August 1994 for the birth of our first grandchild, Macy, and, fortunately, was still there when Seth was mugged on the street in St. Louis. We arranged for him to join us in Niger for Christmas.

Our lives there never settled into a routine. I arrived in Niamey with no assignment, but my earlier efforts soon resulted in a long consultancy with UNICEF that applied my expertise to the country's largely rural and nonliterate environment. The field trips and scores of Nigerien and international colleagues were invigorating. That work led to two more consultancies, one with CARE in Niger, and a second with USAID in West African Guinea. In traditional rural sub-Saharan Africa, being a gray-haired married woman with a grown son, a married daughter, and a grandchild added to my stature and credibility.

I was hitting my stride when, unbelievably, we were in a car accident that splintered my right arm and abruptly curtailed my agenda. Very slowly, I resumed some activities: professional meetings, work with Peace Corps volunteers, and visiting schools and local women's organizations. Now in my early fifties and one of the older women at post, I also hosted get-togethers of women who wanted to do something meaningful. The trials and courage of African women and girls struggling for empowerment (through literacy and resistance to female circumcision) in an authoritarian, patriarchal, polygamous society were a recurrent theme. There were numerous African women professionals, but they all had a story.

In late January 1996 there was a military coup in Niamey. Seth's graduation was in May in St. Louis, and in July the Foreign Service transferred us directly to Dakar, Senegal, for a final two-year assignment (we drove our vehicle to get there, an unforgettable adventure). Then, after three and a half months of settling into a new work environment and our downtown apartment, we took home leave to be there for the birth of Brienne's second child, Cole. Since Seth had settled back in Washington after graduation, we were able to spend the holidays all together again—for the first time since 1992.

Our tour in Senegal was much less stressful. Telephones were more affordable. Besides Gary's work e-mail, I had private e-mail at our apartment, Brienne had an AOL address, and Seth had a work address. Letters and even packages could be sent reliably. Cheaper air travel made it possible for Brienne's whole family to visit and for Seth to bring his girlfriend (now his wife, Michelle) for Christmas. Dakar is a vibrant hub for West African culture and an international venue. Known for our eclectic interests, we were busy. We were befriended by our ambassador and his wife, who were age-mates and generously shared field visits and cultural activities. An international women's book group discussed women's aspirations and the vulnerabilities of personal identities while living in Senegal.

I worked informally with the director of the West African Research Association, Fulbright scholars, and Peace Corps volunteers; made contacts

at UNESCO and other U.N. agencies; was part of an education evaluation team at USAID; and attended conferences. I also spent time writing
and, looking ahead to stateside residence, applied (unsuccessfully) for a research grant and an academic opening.

Gary and I left Dakar with no definite employment prospects but with
a commitment to the Washington area so that we could be near our children and their families (Brienne's son Wyatt was born in 1999, Seth's
daughter, Maureen, in 2006). Again, my networking efforts paid off. That
summer I returned to Africa (Côte d'Ivoire) with a World Bank consulting team on literacy (with two colleagues from Niger and Senegal!). That
fall I taught the graduate seminar in linguistic anthropology at George
Washington University.

In the nine years since my return from living abroad, my professional
life has been active and full. Besides part-time teaching, there have been
five trips back to Africa and six to Europe, plus some new research analyzing videotapes again (with old friends at the Center for Applied Linguistics). There has been some recognition for my work, and I've found a small
niche in the university community. But I've had to work hard at making
that happen, accepting many disappointments and how little I am paid.

Success is always a psychological phenomenon. It's so easy to become
discouraged and dissatisfied with oneself (women are especially prone to
this) and to set aside the work that leads to contribution and personal satisfaction. Too often we accept someone else's judgment or agenda and fail
to nurture ourselves and our dreams.

I've stopped spending my energy applying for jobs and research funding, knowing that decision makers are charged with encouraging younger
scholars and developing institutional alliances. I continue to mentor others myself. It's been sobering to admit that I haven't held a permanent full-
time professional job since my return to graduate school, but gradually I've
become more at ease with who I am and what I've achieved (including
many poems written over the years). I still yearn to complete that book
about language, literacy, and learning in everyday life and the anthol-
ogy on service encounters, but maybe that's too ambitious, and maybe it
doesn't matter. I hope to write more about India and Africa and my life and
work overseas.

My husband continues to be supportive, shouldering the burden of financial demands. Somewhere along the way we noticed turning sixty, and
two years later I had another wake-up call with a serious illness and the
death of my mother. I began to recognize another phase of my life—a truly

senior identity. From that point on, the most urgent thing has been to give, to impart the insights of my own experiences, to make my own contributions, and to help others with theirs.

In this complicated world of multiple identities there is constant struggle for an integrated sense of self and inner strength to inspire good work. The twenty-first century may be another Renaissance. The creativity of women—of mothers especially—is almost in bloom.

Costs and Rewards of Success in Academia, or Bouncing into the Rubber Ceiling

Marla S. McIntosh

Professor of Plant Science and Landscape Architecture, University of Maryland

PhD, Agronomy-Plant Genetics, University of Illinois, Champaign-Urbana, 1978

Born in Chicago in 1952, I came of age during the feminist revolution and began life as a college student at the University of Illinois in 1970. Although I was not too concerned about declaring a major, my academic goal was to *not* study teaching, nursing, social work, or any other girl-friendly major. In my sophomore year I chose a major after my new best buddy, Kevin, suggested forestry. Swept up in the "ban the bra" and the "back to the land" movements, I thought forestry sounded like the perfect subject. I could see myself as a macho forest ranger, sitting in my fire tower, guarding the land. My career goal was to have an outdoor job that did not require a suit or dress. Science had nothing to do with my choice.

Although I liked math because I was good at it and my developing brain appreciated the clear rules and answers, high school science had seemed obtuse and not particularly relevant or interesting. I was the product of a sixties suburban high school where only the traditional sciences were taught, so I chose to study forestry with no clue that forestry included science. This was the extent of my interest in science until I discovered genetics and soil science in college. They were the courses that got me hooked. During the seventies I earned my BS and MS degrees in forestry and PhD in agronomy, and I thoroughly enjoyed my FOW (First and Only Woman) status for each degree. Life was good.

In 1979, my diploma hot off the press, Kevin (my husband as of 1974) and I moved to Maryland. We decided that the first move would be mine since I had an offer to become a tenure-track assistant professor and his offers were limited to postdoctoral positions. Maryland was an attractive location for two scientists because it is home to many universities and federal research centers. We moved knowing that I had a job and Kevin had two postdoc offers in line if their funding came through. Alas, we didn't expect Kevin would be marooned for six months in a bedroom community populated by stay-at-home moms. As time wore on and Kevin was still searching for a postdoctoral position, he became increasingly frustrated, especially since the mortgage for our modest first house left no money to buy a second car or items for him to keep busy fixing up our home and yard. But sometimes when it rains it pours, and one day the phone rang with good news—the grant at Johns Hopkins was funded and Kevin was offered a position. An hour later the phone rang again. It was Georgetown University reporting that their grant was funded and offering Kevin yet another postdoctoral fellowship. Life was good.

Meanwhile, I was a green assistant professor and the only female faculty member in the agronomy department at the University of Maryland. The departmental faculty expressed a genuine enthusiasm for adding a woman faculty member to lighten their teaching and advising workloads. They also delighted in telling me about one elderly professor who repeatedly inquired, "When is that new secretary starting?" Having landed a prized tenure-track position, I felt self-imposed pressures to do everything and be as good as, or better than, the men. My mission was to prove that women can "do it, do it right, and do it with a smile," To me, success would lead to hiring more women for their worth rather than to meet affirmative action quotas. Although I was hired with funds allocated for affirmative action, I was also assured—and I believed—that I would not have been hired unless I filled a niche in the department and had the potential to earn tenure on my own merit. I accepted the job under those circumstances, hoping that with time, affirmative action would become a thing of the past because it was no longer needed.

In my quest to become "one of the guys" and obtain the requisite credentials for tenure, Kevin and I did not even consider having children until my "up or out" tenure year. During that year, my job would be either secured or terminated. What a perfect time to have a baby. Following advice I received from a senior colleague my first week at work, we planned for a summer baby so the birth wouldn't interfere with my teaching responsibilities. Blessed by the goddess of fertility, I gave birth to our daughter in May of 1984, and I didn't miss a single class. That same month, a letter from the University of Maryland arrived notifying me that I had been promoted to associate professor with tenure. Life was good.

My life in the eighties reads like a textbook example of a two-career science couple with kids. We planned our son's birth for June of 1986 so that it, like my daughter's, would not interfere with my teaching responsibilities. Finding excellent day care took a lot of effort but proved essential if I was going to continue a career in academia. In an ideal world, I would have preferred working part-time when my kids were young, but all tenured positions were full-time. Granted, faculty time sheets do not track hours worked, but I was determined to work the same hours as the men.

I was able to work full days with some peace of mind because my kids were in a home environment where they became playmates of their day care provider's daughters. Juggling schedules around sick and cranky children was more problematic, and I plead guilty to taking my children to day care at times when they complained of being sick. When the kids qualified as really sick (had a fever), Kevin usually "volunteered" to stay home with them. He was the "real" scientist and could work in the lab at night, whereas I needed to be at work during the day to teach class and conduct field research.

Kevin and I shared picking-up and dropping-off responsibilities, which were often the worst parts of our day. Getting the kids dressed in the morning was particularly arduous until we decided to have them choose their clothes the night before so they could sleep in them. One year, I bought our son Superman pajamas that he slept in and then wore to school for Halloween. It was amazing how much life improved when he could roll out of bed and sleepwalk to the car.

Although it may sound cold-hearted, I did not want to be at home with kids 24/7, and I selfishly chose to leave them each day while I escaped the world of babies and toddlers for the stimulation of the adult world. I rationalize that, because of who I am and who they have become, my family gained more than they lost by having a working scientist mom. I think it would have been better all around, however, if I had been able to work part-time during those demanding years.

By all appearances, I was successful. While balancing career and family, I was promoted to full professor and later associate dean for the College of Agriculture and Natural Resources. Until becoming associate dean, I rarely interacted with other women faculty, since women within the College of Agriculture were still few and far between. When I became an administrator, however, my perspective changed, as did my concept of success. I began to notice pervasive family-unfriendly attitudes that negatively affected retention and job satisfaction among women faculty. It gradually became apparent that expecting faculty to work long hours and weekends at the expense of their personal lives did not increase productivity. The pressures to work late in the lab because you were a researcher or bring papers home to write and read because you were a scholar and teacher took a toll on faculty, particularly women with children. Younger women in my college confided in me that the quest for tenure wasn't worth it. Their priority was family, and regardless of the enormous benefits of a faculty position, they did not want a job where working *only* a forty-hour workweek was perceived as lacking dedication and commitment.

Until the mid-nineties I accepted that traditional macho work ethic, but no longer do I accept it as the way things should be done. While in the inner circle of college administration, I tried working from within to bring change but succeeded only in becoming increasingly marginalized. There was a cascade effect, and my opinions were summarily dismissed before I could even express them. Superficially, I had an image and title at the university that was associated with rank and power. In reality, I had hit a "rubber ceiling" where I repeatedly worked my way up only to be bounced back to the floor.

In 2000 I resigned as associate dean and e-mailed the college faculty:

I strongly believe that it is time for the college to move to a new phase of growth and increasing excellence. In order to accomplish this, Academic Programs needs to rise above its present status as the weak sister in the College and receive a commitment of additional resources. I have tried to lead Academic Programs with limited staffing and discretionary budget which has hampered my effectiveness to accomplish the myriad of activities needed to bring Academic Programs to the level of excellence that it is poised to reach. Therefore, I feel that it is time for the College to seek a new Associate Dean of Academic Programs who can be a more successful advocate for our students, our instructional faculty, and their support staff.

I look forward to resuming my position as Professor of Natural Resources Sciences and Landscape Architecture and Distinguished Scholar-Teacher to focus again on teaching and research. I continue to be committed to contribute to building a great university, but in a different way, a way that should be mutually beneficial to myself as well as the University.

My story is about a woman in science whose life has been good and whose expectations were met or surpassed. The broader theme, however, is that my expectations were framed by my generation, my personality, and my gender. The same life choices might not have been as good for a woman of a different generation or with a different personality. In hindsight, I realize that the "build it and they will come" premise that motivated the women of my generation did not work. The percentage of tenure-track faculty in colleges of agriculture who are women remains the lowest among the life sciences.[1] In my own college the number of female full professors is down to three. What happened to the cohort of women hired in the wake of the feminist revolution? Could it be that, when confronted with choices pitting career against family, they chose family and either dropped out or were not promoted?

My career has been intellectually rewarding, providing the satisfaction of seeing positive impacts from teaching and research and affording opportunities to travel around the world. But while the path to success has changed in other fields in response to a changing society, *academic* success clings to traditions that are out of step with the world beyond the ivory tower. Thus, the rewards and the barriers to motherhood remain constant.

In summary, there is no way that I can describe how my children have affected my life because they are an intrinsic part of it. At first I chose my career, but once born, my children refocused my life and changed me in unpredictable and incalculable ways. I can't, nor do I want, to imagine my life without them. Oh, and did I mention my husband?

Once again, life is good.

1. J. S. Long, ed. *From Scarcity to Visibility: Gender Differences in the Careers of Doctoral Scientists and Engineers* (Washington, DC: National Academies Press, 2001), 67, 70, Table 4-4.

One Set of Choices as a Mom and Scientist

Suzanne Epstein

Immunologist, FDA Center for Biologics Evaluation and Research

PhD, Biology, Massachusetts Institute of Technology, 1979

My first and most important science mentor was my father, a chemist, who occasionally took my brother and me to visit his laboratory, worked on projects with us at home, and read to us books like Rachel Carson's *The Sea around Us*. When he reached the end of that particular book, he told us with a twinkle in his eye that there would be an oral exam. There was only one question: what topics about the sea were not included in the book? Our discussion of that wonderful question went on for a long time.

Other early mentors were a beloved nature counselor at camp and a dynamic high school biology teacher, both women. In college, I had an inspiring man as my freshman chemistry teacher, full of energy and enthusiasm. I became one of very few female chemistry majors at a time when there were no female faculty in the department.

Toward the end of college I became frustrated with what I saw as the ivory tower type of science. An uncle thought I should go to medical school. His theory was that an MD would allow a woman to reenter the professional world, after a hiatus to raise a family, with more power and prestige than a PhD. However, I was interested in scientific questions in a way that premedical students were not. At the time, I knew that if I were to leave science I would enter a completely different field, and if I were to stay in science it would not be in medicine.

I had applied to some biology and biochemistry graduate programs and had been accepted but was not yet ready to commit to graduate school. I applied for and received a fellowship to spend a year in Germany. Deferral of graduate school for a year was not an option back then, and all six schools informed me that they were throwing away my materials. I would have to reapply from Germany and provide new transcripts and letters of recommendation. On top of that, my chemistry advisor said he was sorry to hear of my plans because, as he put it, I would never be readmitted and would thus end my career in science.

Despite those narrow views, I decided to go. It was a marvelous opportunity to experience another culture, not just as a tourist but as a resident. I spent the year in Munich studying a variety of subjects, making friends, playing music, and traveling. I made many visits to the post office to send graduate school applications, and despite my advisor's predictions I was accepted again. Meanwhile, I decided that the type of science that would hold my interest would be a socially relevant field. Environmental engineering and nutrition came to mind, but I had not applied to any such program. I entered a graduate program in biology at MIT with the intention of transferring to another department if I did not like the biology program. During a series of faculty talks to acquaint students with their research fields, I discovered immunology and felt I had found my niche. This science supported the development of vaccines that benefited people in all walks of life.

During my fourth year of graduate school I was supported by a fellowship for women that specified that I was not allowed to marry, presumably to encourage women to complete their studies rather than drop out. I was living with my boyfriend at the time. I chuckled at the rule but figured I was fulfilling the wishes of the benefactor (Alice Freeman Palmer) because I was completing my thesis research. After the fellowship ended, we decided to marry. It was a female biology professor, not a man, who upon hearing of my plans to marry came and told me that she was sorry to hear I was dropping out of graduate school! I told her I certainly was not. I guess she couldn't imagine the combination.

She was no role model, and throughout my career I have had to ignore many other unsuitable examples of scientists, both men and women, whose priorities I did not share. A college friend with children turned out to be my best role model. At one point a major university was recruiting her. When negotiations progressed and they offered her the job, only then did she tell them that she wanted to work part-time. They were surprised but permitted it. Slower tenure clocks for part-time faculty were not offered back then, so she proposed and obtained an equivalent option. She has had a highly productive career and is now a full professor. Though her career has had its problems and frustrations, it was designed as much as possible on her own terms. The lesson is not in the specific options she chose but in the fact that she thought for herself.

Following graduate school I went on to a postdoctoral research position. My husband and I wanted to have children, and my view of commitment to my new job was that I would give at least a solid year of work before starting a family. When I eventually did become pregnant, I told my boss that I wanted to take three months of maternity leave and then return part-time by working five shorter days. His reaction, quite unexpectedly, was that he evaluated performance not by the number of hours per day but by progress over the course of the year. Later that year he nominated me for promotion. Meanwhile, I had to take maternity leave without pay and pay cash for my health insurance, but I felt fortunate to keep my job in research. My husband also took three months off without pay. His company had no paternity leave policy and when he asked, they didn't know what to say. They passed the question up the chain of command, finally reaching a woman in another city who said yes. Fortunately, we had enough money saved to make our time off possible.

The three months at home with our son were a glorious and exhausting upheaval. I enjoyed the time with both him and my husband. I nursed my son, played with him, and learned to pump milk (after all, I am an immunologist). I searched for home day care as a way to expose our son to fewer people's germs and to provide a single, chosen caregiver rather than the changing staff in a large center. I went back to work with great uncertainty. During my first day back I worried about how my son would do at day care, figuring that if this didn't work out, I would switch to a less demanding line of work in order to have enough time at home. That evening he rolled over and looked very proud, as if to say, "Look what I learned today." I realized that my children would guide me as to whether our approach was working for them as individuals. My husband and I settled on the pattern that I would stay home later in the morning, while he would get home first after work. That way I had the flexibility I needed at the end of the day, as my experiments were less predictable than his work. Once, when my mom visited my son's day care home, she said she wished there had been options like that when she was raising us, because we would have benefited greatly from the interactions with other children. At day care my son had the wonderful experience of getting to know some special-needs children in the group, an experience that he might not have had otherwise.

Until I knew how the combination of family and science would work, I had hesitated to establish a lab of my own. A couple of years later, when someone I knew recruited me for a position at the Food and Drug Administration's (FDA's) Center for Biologics Evaluation and Research, my son was thriving and I was ready. The job was an option I hadn't known about: a combination of research and practical public health work in a government agency. After other interviews and deciding against a couple of alternative offers at the National Institutes of Health, I accepted that job and had our second son after working there for a year and a half. Again, I took three months off, and this time my husband took two months. While pregnant, I had a site visit review and was nominated for tenure. In an unexpected twist, I was on maternity leave when my boss called me at home to say I had just received tenure. Once again I worked five shorter days for the first year. This time hours were counted strictly, so after using my available leave, I used leave without pay. Again I pumped milk; both our boys were very sensitive to soy or dairy, even in my diet, so breast milk was their only milk for the first year, and they nursed for quite a while after that. Again our baby went to a small home day care group.

A female scientist I encountered at work some time later struck up a conversation about children and asked how long a maternity leave I had had. When I said three months, her jaw dropped. She'd had only two weeks. "How did you ever get that much?" she asked. I said simply, "I asked. Did you?" The answer was no. Perhaps she was influenced by a very competitive career environment, and her fears had denied her an option she claimed she would have liked.

Our family was lucky to live close to my job and to the children's schools. I could quickly get to school events or pick up a sick child. I was also fortunate to have warm relations with neighbors. Three of us on the block were mothers of young children at the same time, and when I was working full-time, one of the others worked part-time and one was at home. Our differences in approach to career and family were not an issue to any of us, and we are still friends.

I felt that parents who wanted more options should help create them, which is why I served for many years on an NIH campus day care board that gave input for enlarging and enhancing sponsored programs. A satisfying personal accomplishment was negotiating the addition of a basketball court and a tennis court for the school-age kids at one center.

After receiving tenure, I changed fields to work on something far more interesting to me than the type of immunology my mentors worked on. By apprenticing myself in other labs to learn new techniques, I moved into the field of viral immunity. I began research on immunity to influenza virus infection, with the goal of better informing vaccine development. This type of practical work was looked down on by some colleagues, and the influenza system was considered passé in the early 1990s. Nonetheless, I pursued my interest and established myself in this field. I find it ironic that now, suddenly, many people understand the importance of this research.

At various times after establishing my own lab, I have had to temper my ambitions for progress in research in order to accommodate the needs of my postdocs and technicians, male or female, who had family duties, including caring for a special-needs child or a sick or injured child or parent. Female scientists who succeed in establishing themselves and becoming mentors have the same responsibility as any scientist to respect the needs of their staff.

Our sons are now grown. One has graduated from college and the other is in college. The relatives who were skeptical of our choice to send them to day care so young have long since come around, and admire the young men they have become.

What does this add up to? It's true that I was lucky to have humane bosses, but I made an effort to choose them. Women I knew who were ambitious above all else might not have selected those jobs. I also had to accept the consequences of my choices, such as not traveling a lot or having a large lab. My chosen career path in a nonacademic setting was also looked down on by some. A few years ago my graduate school sent a letter inviting alumni to give talks on career options, and I volunteered. I never received a reply.

Young scientists, both men and women, make decisions about priorities at a vulnerable time, when their own views are still developing and their careers are not yet established. Powerful senior scientists have control and make decisions that could adversely affect careers. However, they do not have better judgment about the lives of others and are not necessarily better scientists. Young scientists must choose their own goals and their own approaches to career and family life.

SECTION II

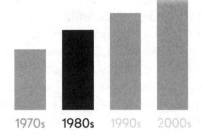

1970s **1980s** 1990s 2000s

SINGLE-MINDED DEVOTION

It is dead silent in the laboratory at 3:00 a.m. I fear that I am the only one in Haring Hall, the cavernous old building on the University of California, Davis campus. The −80 °C freezer where I store my precious samples of mouse lungs, livers, and spleens is in the basement, three floors below. My mother would have a fit if she knew, and I know I will never tell her that I am here alone spinning samples on a campus where the partially completed 46,000 square-foot veterinary diagnostic lab has just burned to the ground, the letters ALF (Animal Liberation Front) painted near the fire. But I am driven by curiosity about the data I will collect, the analysis I will perform, and the results that will follow. It is 1988, the final year of my PhD, and not once since discovering the field of toxicology have I questioned my career choice.

· · ·

By the early 1980s, in contrast with the seventies, women were practically pouring into science studies. Many previously all-male schools, including those focused on science and engineering, were now open to women, and the women of the 1980s came in force, no longer pioneers. Three women won Nobel Prizes in the sciences in the eighties—more than in any other decade before or after—and female faculty at the Massachusetts Institute of Technology jumped from seventeen in the seventies to seventy-one in the eighties. In 1983 Dr. Sally Ride became the first American woman in space on the space shuttle Challenger, but in 1986 the shuttle disintegrated seconds after takeoff with Christa McAuliffe, mother of two and the first representative of the Teachers in Space program, on board.

Numbers of women undergraduates in computer sciences and engineering doubled and tripled, while research and development expenditures at the top research universities indicated that "women accounted for all net growth in science faculty at the assistant professor rank."[1] In 1989 *Science* reported that the NSF predicted a shortfall of 675,000 scientists and engineers by the year 2006. The answer to that shortage? According to the same article, "all are agreed: If a long-term shortage of scientists and engineers is to be averted unprecedented numbers of women and minorities will have to be attracted to technical careers."[2]

Women earning their PhDs in science in the 1980s were no longer aberrations; they had not been encouraged to stop their studies, get married, and have children. They were considered an integral part of the science and engineering profession in the United States. These members of the "career *and* family" generation of women were encouraged to reach for the stars both at home and at work.[3]

1. J. T. Bruer, "Women in Science: Lack of Full Participation," *Science* 221 (1983): 1339.

2. C. Holden, "Wanted: 675,000 Future Scientists and Engineers," *Science* 244 (1989): 1536–37.

3. Claudia Goldin, "The Long Road to the Fast Track: Career and Family," *ANNALS, AAPSS* 596 (2004): 20–35.

Three Sides of the Balance

Anne Douglass

Atmospheric Chemist, NASA

PhD, Iowa State University, 1980

Much has been written in the past few decades about women's efforts to balance work and family, including the challenges facing women in science. The balance between family and work can seem precarious when mommy is a scientist. My daughters and I think we are unusual, maybe unique. I have survived the balance and created a successful career in atmospheric science; my five children have grown and thrived; and two of my daughters have chosen scientific careers of their own. Katherine, my oldest daughter, is a physician specializing in emergency medicine. Elizabeth, my second daughter, is about to complete her graduate studies in oceanography. My essay tells of my journey from my bachelor's degree in 1971 to my present position as a NASA senior scientist with grown children. Later in this book, my daughters describe their sides of the balance. They offer thoughts on growing up—how my balance worked for them—and on the challenges that they face today.

I don't think the question is whether one can be a mother and have a career. Both are possible; lots of people do it. The question really is, can a woman have a productive career as a scientist and be a *good* mother? I think that I have done both. Love of family and love of science played an important part in getting me to the present. A combination of pragmatic decisions, practical accommodations, luck, and a lesson learned also figure prominently. Before we get to these, here are some facts. In 1971 I received a BA in physics and became a mother. In 2005 my youngest child, the last of five, graduated from college. In the interim, I completed my PhD, went to work for NASA, wrote about a hundred papers, was elected Fellow of the American Meteorological Society and Fellow of the American Geophysical Union, and did lots of other things that NASA scientists do. Today my family obligations have been mostly replaced by mutually fulfilling relationships with young adults and their spouses and significant others and our lives are sparkled by five (six years old and under!) grandchildren. I am engaged in chemistry/climate modeling and am one of the project scientists for Aura, a major Earth-observing satellite. It's a logical time to reflect on my personal mishmash of family life and scientific endeavor.

My first love is family. I always saw myself as a mother, and while I did not plan a five-child family, I cannot imagine my life without each of them. Children bring you joy, children demand time, and above all, I think children need unconditional love. I fell totally in love with my kids and from that love drew the energy I needed to take care of them. I worked about three-quarter time in graduate school and also for my first ten years at NASA because I couldn't fit in all the necessities (let alone the fun) if I tried to have the career full-time. There were lots of times this felt like a sacrifice, but it almost always felt worth it.

My second love is science. I graduated from high school expecting to major in math, teach high school, and get married. Physics was required for a math major at my college, and I found myself enchanted by the patterns, the connections, the sheer beauty of physics. Learning more has opened doors to other things I want to understand. I focused totally on work while at work and mostly on family while at home, but sometimes when engaged in mundane activities like washing and carpooling, I had ideas floating in the back of my mind. Often I would leave work with a problem and come back the next day with the solution or at least an approach to one—never on paper, always in my mind.

I had made some important decisions by the time I completed graduate school. I started with the idea of earning an MS in physics, taking the plan B "no research" option. This idea lasted less than a year. The attraction of research and the confidence that I was PhD-worthy overwhelmed my prior simple career goals. My research advisor at the University of Minnesota first allowed a part-time lab schedule and then made space in the lab for a baby carriage in the corner so that I could continue to breast-feed my infant son. I was the only female graduate student in my cohort, and the accommodations seemed reasonable to me at the time, but looking back, I can only imagine what some of my male fellow students (not to mention the faculty) were thinking. Through my lab experience I gained insight into vibrational energy transfer and chemical kinetics. I also gained practical experience balancing the unpredictable requirements of finicky experiments with the predictable *and* unpredictable needs of preschool boys.

When my husband finished his PhD on the fast track and took a faculty position at Drake University in Des Moines, Iowa, I had to figure out how to balance my graduate career and my growing family while adding a forty-mile commute to Iowa State University in Ames. I made a choice that was both pragmatic and idealistic to focus on atmospheric science. I left the lab forever (pragmatic), substituting satellite observations for data I had obtained myself. I wanted my research to contribute to the public good and the future world for my children (idealistic), and so I chose earth science and the multidisciplinary world of global change research. My modeling and analysis work continues to fascinate and challenge me, grounded in the truth learned from remote sensing. Computer simulations and gigabytes of satellite data are far more portable and far less temporally demanding than liquid nitrogen dewars. I contribute to World Meteorological Organization assessments of change in stratospheric ozone and to Intergovernmental Panel on Climate Change activities. I honor this responsibility.

When I finished my PhD, I applied for faculty positions. I was forthright about my family (then two boys and two girls) but thought that my awards as outstanding teaching assistant and my experience teaching classes on my own would make me a good candidate. I was never offered an interview, however, and so I took a research position at NASA Goddard Space Flight Center (GSFC). This was a fortunate occurrence, as GSFC offered practical accommodations that were then nonexistent in academia. After ten years I converted to full-time, when my oldest started college and my youngest started first grade. On-site, high-quality day care was a key to my success—so much so that my bosses and friends knew that I would never seek a position elsewhere as long as I had a child in the day care center. For my entire career at GSFC I have had two kinds of flexibility. First, I could exploit "flex-time" to its fullest, working nights and weekends when sick children or other family issues kept me away during normal working hours. Second, for the most part, when my children were young my research was not time-sensitive; model development, application, and analysis can progress for weeks or months with few external deadlines.

For me, the burst of scientific research accompanying the discovery of the Antarctic ozone hole was a piece of luck. Suddenly there was more work to do than anyone had dreamed possible only a few months before. Although I was never part of the forefront of polar science, I was able to make contributions through modeling and data analysis on longer time scales. It is easier to advance in an expanding field than in one that is stationary or contracting with fierce competition for research dollars. Because I was able to make several significant contributions, I was offered a leadership role in a NASA satellite program in 1993. I was able to accept that offer because, when it came, three of my children were college age. This established my career at NASA.

The lesson I learned is patience. There were times when I thought they would never grow up. When my second daughter was a baby and all my graduate school friends were driving to North Dakota in February to see a total solar eclipse and I had to stay home because I was afraid it would be too cold for the baby (it turned out to be $-22\ °F$), I wasted mournful time thinking I would probably die before the children were grown and everyone would say, "She had a lot of potential." When my last daughter was little, I watched my colleagues go off to meetings to present my work and thought that I would be a support person for my whole career and never get to go anywhere. The children do grow up, though, and since then I have had opportunities to be a group leader, to take my ideas to scientific meetings, and to speak at universities in the United States and abroad. The opportunities are there because I was able to lay a strong foundation for my work, quietly, while I was raising the kids. Child rearing has even turned out to be a useful experience when it comes to organizing a research group, a scientific meeting, or a science team. In the same way that "balancing" a family taught me things that I could bring to my professional life, being "balanced" has empowered my daughters as you can read in their essays (Katherine Douglass, MD and Elizabeth Douglass, Graduate Student). If I could give you a gift, it would be the patience to recognize that childhood is precious and fleeting and that science will be waiting for you with some awesome mysteries when your children become adults.

The Accidental Astronomer

Stefi Baum

Director of the Chester F. Carlson Center for Imaging Science,
Rochester Institute of Technology

PhD, Astronomy, University of Maryland, 1987

Graduating from college in 1980, I was determined to have a large family (four kids, dogs, cats, the whole nine yards) in addition to a career. I was married in graduate school to another young astronomer, and my husband and I have made many decisions guided by our "family first" priority, sometimes putting our careers at risk at critical stages. This could have turned out badly, but in the end, probably with a large amount of luck and certainly with an enormous amount of hard work (both in the lab and in the home), we persevered. For me, the important lessons are (1) do what you love, (2) go after what you want, and (3) stick to your priorities.

Looking back, it's hard to believe that I wound up where I did. As a child, I always wanted to have a big family. I was also curious about science and enjoyed using my brain. I loved to read fantasy books where people, animals, and creatures did noble and inspiring things. My father was a mathematician and my mother a high school teacher; they pushed me academically and expected me to succeed, but since I was dyslexic, I was only a B student. I was very good at sports and enjoyed them. Team sports, in particular, appealed to my sense of a noble conquest, and this desire to be part of team, developed early in life, has been a hallmark of my career.

Unusually high ("for a girl" and at that time!) SAT scores and advanced coursework in mathematics helped open college doors for me. I entered college at Harvard and started as a math major, but I kept exploring. I changed my major to biochemistry, then to chemistry, and finally to physics. I was the only girl in many of my classes. My advisor once asked me why I was majoring in physics since I would never make it through graduate school, and if I did I would never be an actual "physicist." Assuming that my esteemed professor was right, I gave up plans for graduate school. I was heavily involved in sports in college and was cocaptain of my lacrosse and eastern championship soccer teams. Learning how teams work turned out to be important later in life when I found myself repeatedly thrust into management and leadership roles. My parents, though obviously proud of my accomplishments, were concerned when I graduated from Harvard without finding a husband!

Immediately after graduation I accepted a job teaching chemistry in Israel, but the job fell through. After searching laboratories across the Harvard campus I was hired as a data aide at the Harvard-Smithsonian Center for Astrophysics in Cambridge, Massachusetts. Thus began my long career in astronomy. At the Smithsonian I worked under a married couple, Christine Jones and Bill Forman. I liked astronomy, and I loved working for Christine and Bill. In addition, they had a dog and were about to have a baby. They seemed to have the things I wanted. When an astronomy professor at the University of New Mexico called Christine and Bill looking for a graduate student, they recommended me. The following January I moved to Albuquerque and started graduate school.

The next momentous occasion was meeting a handsome young man who was also an astronomy graduate student doing his PhD research at the Very Large Array radio telescope in New Mexico. He smiled at me an enormous frog smile, and we fell in love. We decided that we would not let our careers get in the way of our relationship. Chris was a couple of years ahead of me, and when he finished his PhD he took a postdoc in Charlottesville, Virginia, and I transferred to University of Maryland. Chris turned down a faculty position at University of New Mexico since it didn't seem that there would be a permanent position for me. I arranged to do my PhD thesis research at the National Radio Astronomy Observatory (NRAO) in Charlottesville. My advisor at Maryland was very supportive, as was my NRAO advisor. I had coadvisors in Holland and California. All of a sudden I felt I had entered a warm family of astronomical research and I was off and running scientifically. Chris and I got married in August 1985 in our backyard. Our toast was "The Adventure Continues," which has become the theme of our lives.

Just before I received my PhD we were both awarded postdoctoral positions at the Netherlands Foundation for Radio Astronomy, in a small village called Dwingeloo. It was a dream come true. Chris turned down another faculty position so we could take the postdoc jobs together. I defended my dissertation while seven months pregnant. We immediately left for Holland, beginning new jobs and having our first child in the space of a few months. When we arrived in Dwingeloo and the head of the astronomy group saw that I was pregnant, he groaned. I had not informed them of my pregnancy.

Giving birth in Dwingeloo was an experience. There was no hospital in the village, and our first child, Connor, was delivered in our bedroom by Dr. Dinkla. I went back to work one week after Connor's birth. At first we took Connor to work with us. The Dutch men would laugh at Chris when he carried Connor around in a pouch. Soon Connor was too old to take to work, and we found a local woman, Clara Looman, to take care of him during the day. We became good friends with her family and learned to speak Dutch. We had decided we wanted our children to be close together in age so they would interact well together. Eighteen months later our second child, Kieran, was born.

After three years we left Holland and moved to Baltimore. Judith, the daughter of the Looman family, came to the States with us and served as our au pair for our first year back, making the transition much easier for us and the kids. I was pregnant at the time, and Brennan, our third son, was born five months later. I took a Hubble Fellowship at Johns Hopkins University, and Chris started a tenure-track position (finally) at the Space Telescope Science Institute (STScI) across the street. A year later I took a faculty position at STScI. Our "two body problem" had been solved. When I was interviewing for the job at STScI, I was many months pregnant with our fourth child, and one of my favorite moments was when astronomer Meg Urry stood up before my seminar to announce me, and she, too, was very pregnant. We were quite a pair and a tribute to how far women in science had come. Annelies, our first girl, was born a few months later, joining her brothers in the world. I was thirty-four years old, we had four kids spaced regularly eighteen months apart, and our family was complete.

But life was exhausting. We had scientific careers to maintain, proposals to write, discoveries to make, papers to publish, functional work to accomplish, tenure to be won, and a teeming family of active kids spilling everywhere we went. It was a wonderful and sometimes overwhelming period of our lives. I was up at four most mornings, using the wee hours before the family woke up to do all those things, for work and for home. After the kids went to bed, Chris and I would stay up to talk science and squeeze a little more work into the day. To help in the home, every year we had a new au pair from a different country. Many of the au pairs became lifelong friends. They added to our children's education, cultural awareness, and perspective. From the beginning, Chris and I shared in the household responsibilities and child rearing. When our kids were babies, I took more responsibility, but as they got older, Chris assumed a greater and greater share, and today he does more than I do. We didn't have rules or a prescription or measure who was doing more; we just approached our family, our scientific careers, and our lives as a team. We collaborate on both scientific research and raising our family. We became specialists with different functions in the home. I haven't written a check or paid a bill in more than twenty years, and Chris rarely, if ever, cooks breakfast or dinner.

At STScI, astronomers spend half their time doing research and half doing "functional work" in support of the Hubble Space Telescope or the James Webb Space Telescope. I continued my research on the nature of activity in galaxies and started working to develop the first truly functional scientific archive for a major observatory. It was a team environment and I loved it, working with scientists and engineers to create something new. With that accomplishment, I moved on to join the instrument team supporting a new instrument to be placed on Hubble, the Space Telescope Imaging Spectrograph (STIS), and soon became lead of the STIS team. It was a work-crazy period of my life, but it was a thrill and an honor to be able to help bring such an amazing scientific resource into the hands of the community. We took the whole family down to watch the launch of the shuttle that carried STIS into orbit.

In 1997 my mother became gravely ill. Chris and I were planning to take the family back to Holland for our sabbatical year, but we quickly changed our plans, taking our sabbatical in Princeton so I could be with my mother and father there. My mother lived four months, dying from cancer that spring. My kids were with her in the last months of her life. It was a hard time, but we were glad to have been able to show the kids what family is all about and to have let them experience a "good death," if there is such a thing. That spring, just weeks after my mother's death, our dog gave birth to a litter of twelve puppies. We kept the one that looked as if she had a ring of pearls around her neck and named her Julia in honor of my mother.

After returning from sabbatical to STScI, I became head of the engineering division, responsible for support of the Hubble ground systems and supervising 140 engineers, scientists, and support staff. It was an enormous challenge and one I was afraid I wouldn't be up to, but I found I loved the job—it was another noble team effort to take on. I found I had a calling for and love of coaching, organizing, and mentoring an organization like this.

STScI was a male-dominated environment when I first arrived there. I remember clearly how all astronomers were spoken of as "he" and never "she." And there were no family leave policies or tenure-clock-stop policies at the time to support young scientists and engineers as they started families. But change was under way. Under the leadership of Riccardo Giaconni (who later won a Nobel Prize), and helped along by new federal laws covering sexual harassment, equal pay, and so forth, a new culture was coming, albeit slowly, to STScI. A fellow scientist, Hal Weaver, and I created an ad hoc benefits committee, and we worked with the senior management of STScI to create family-friendly leave and benefits policies. Under the drive of Meg Urry and Ethan Schreier, STScI hosted a meeting on Women in Astronomy and created the Baltimore Charter to support the advancement of women in the field. But culture is a hard thing to change, and while the law can remove the monstrous inequities, it cannot remove those many microinequities in terms of how women are treated, respected, and viewed in the world of physics. The small number of women astronomers at STScI seventeen years after I first went there is an indication of how long and hard you have to push to turn a culture around.

Over the years our kids continued to grow and flourish, entering grade school, playing sports, trying (and failing) to learn to play instruments, and enjoying life in the neighborhood with their friends. When the kids were babies we sometimes took them with us to meetings, but as they entered school, I stopped traveling. Recently, as the kids became teenagers and more independent, I began to travel again. Still, I haven't been to an observing run at the major national and international telescopes used for my research for more than eighteen years, and I only rarely attend scientific meetings. It has been a career sacrifice worth making.

After a couple more years I decided it was time for a change and took an American Institute of Physics Fellowship at the U.S. State Department, where I served as a Science Policy Fellow for a year and a half, focusing on issues involving genetically modified organisms and agricultural biotechnology. All those years in which I criticized our government from my position in the ivory tower dissolved away as I experienced firsthand the dedication, work ethic, and ability of the folks working on our nation's (and the world's) behalf.

After fourteen years in Baltimore, Chris and I, for a host of reasons, felt it was time to move on from STScI and try our hand in academia. We got jobs at the Rochester Institute of Technology in Rochester, New York. Chris became a professor in the physics department, and I became a professor and Director of the Chester F. Carlson Center for Imaging Science. We agonized over the impact a move would have on our kids. Annelies and Brennan were in middle school and Connor and Kieran in high school. Removing them from their environment and friends would be a jarring change. The funny thing is that as good as the move has been for Chris and me (we just love the involvement with students and the freedom being in academia provides), the move was even better for our kids. They have flourished in this environment, making new friends, rediscovering themselves in new ways, realizing their academic abilities, and excelling in sports. What more could we ask for?

After writing this essay and talking about all this with Chris, we agree we wouldn't hesitate to do it again, but we also realize that we wouldn't have to anymore. We recently had a postdoc working with us who decided to go back to Australia to take some time off from research and have a baby. She has been able to win a "career interruption" fellowship designed to help her return to research. To think that I was back to work only one week after having my first child and that I timed my pregnancies so as not to be visibly pregnant when applying for my first jobs. We were crazy, and the times were crazy. Change comes slowly, but it does come, and it is so good to see. We should all keep pushing for equity and respect for all who enter the hard science and engineering fields and for the creation of sensible career paths so that the young men and women who accept the challenge can do so in ways that give them plenty of time to enjoy both family and research along the way. The adventure most certainly continues for us all.

At Home with Toxicology

A Career Evolves

Emily Monosson

Environmental Toxicologist and Writer

PhD, Biochemical Toxicology, Cornell University, 1988

A few years ago I attended the twenty-third annual meeting of the Society of Toxicology and Chemistry (SETAC), an organization to which I'd paid my dues for more than ten years. In organizations like SETAC your affiliation, be it academia, government, nongovernmental organization, or industry, is your badge of honor. It's also required by the computer that spits out the badges. So when asked for my affiliation, to the horror of my traveling companion and colleague at the time, I said "housewife."

Admittedly, I was feeling particularly accomplished that day, so I didn't mind "outing" myself as a part-time scientist. Besides the family, I was juggling a consulting job, a research project, and a class that I taught each year at one of the local colleges. I could do it all and be at home after school for the kids, bugging them to eat an apple or do their math homework.

What I did not expect when I squeezed my way through the crowds or waited patiently in line to ask questions of poster presenters, was that by revealing my lack of affiliation, I'd become invisible. The minute the scientists and graduate students who were milling about or attending to their posters read my badge, their eyes glazed over, or worse, they looked over my shoulder to the next in line. I was no longer worthy of their time. I was, after all, just a housewife. What could I know about science?

That badge still hangs on the bulletin board beside my desk. It's the only one I've saved after fifteen years of attending all sorts of scientific meetings. Why? I certainly don't need it to remind me of my multiple roles in life. But it does remind me of a time when I was not embarrassed to announce professionally—to all whom I'd previously only presented my scientist self—that I was sharing my career with my family.

That was three years ago. Then I hit one of those all-too-common career troughs suffered by many of us who try to make our own way. Generally it happens every two to three years, depending on the grant cycle. But this one was different. I'd just turned forty-five, the kids were older, and though it was still important that I be there for them after school, they were more independent and I had more time. I realized that I was ready for more: more work, a more stable career. But having been out of the mainstream and essentially unaffiliated with any particular organization for so long, I began to wonder if my family-friendly career choices had resulted in an inadvertent detour from a sustainable career.

As a girl, before I'd ever heard of the word "toxicologist," I had played mad scientist, mixing potions with the cleaners and solvents stored under basement sinks (the ones we all keep locked away these days). With friends, I'd test the potions on earwigs and carpenter ants that infested the old wooden jungle gyms behind our homes or raced along cracks in the old concrete patios. Only once, when a small dish of the stuff we had stashed under the couch evaporated overnight, did we fear that my friend's younger sister or, worse, her pet cocker spaniel, Buffy, had licked it up.

Throughout high school and college I was never at risk for dropping out of the sciences; they were what I did best. As a graduate student at Cornell University, I rediscovered my love of toxicology, and several years later as a postdoc I discovered another love—my husband, Ben, a graduate student in ecology at the time.

At that point my parents—who, although they never really understood my world of science, did understand my devotion—voiced their concern. "Never let a man tell you what to do," they said, worried that I might throw away all that work. The fact that they'd never met Ben until after he became my fiancé likely added to their concern, but it was not the first time I'd been warned about mixing my love life with my science life. Over beers one night my undergraduate advisor, and a colleague of hers gave me some advice for graduate school and beyond: Don't marry another scientist. Although she and her husband were (and still are) happily married and both were faculty members in the same department, she had held a "visiting faculty" position for fourteen years, while her husband advanced along the traditional tenure-track path. Recently I asked her about this. She wrote, "When we came to Union in 1973 my first kid was nine months old. I wasn't job hunting and had my head in the sand about my future—but you know what it's like to be a first-time mother. Besides, at that time no one was interested in helping 'the spouse' in any serious way." Several years ago, after more than twenty years at the college, she was finally promoted to full professor.

When I was a graduate student, long-term career decisions seemed years away. Besides, the only men I met were those who shared the same −80 °C freezers, the same vending machines, the same computer printers, and the same wet labs. In fact, because I never really understood the dynamics of water flow, often flooding the wet-lab floor with potentially toxic water as I drained too many experimental tanks at one time, I got to know Ben as I mopped up the floor surrounding his own nontoxic experiments. So began my career as a scientific vagabond.

When I met Ben, I held a National Research Council postdoctoral position with the U.S. Environmental Protection Agency (EPA) and hoped one day to convert to a full-time position as an aquatic toxicologist. I was ready to settle down and build a career, but Ben was just completing his master's and moving on to a PhD program in North Carolina. Confident that I could make it work and unwilling to carry on a relationship states apart, I applied for and was granted funding to do research down in North Carolina. Ever since then, and for the past fifteen years of our marriage, my career has been a continuous stream of one- and two-year stints on one project after another.

Shortly after he completed his degree, the U.S. Geological Service (USGS) offered Ben a job as an ecologist in the "Happy Valley" of western Massachusetts, two hours from my family, a bit more from his, and within an hour of five colleges. We jumped at the opportunity. By that time, we'd spent a year in Stony Brook, Long Island, where our son Sam was born and where a colleague and I had just finished the first of two years' planned research; the second year could easily be done elsewhere. Eager to pull one-year-old Sam from the industrial-sized day care where he'd spent six months (six months of ear infections and antibiotics), we joined a family day care just a block from our new home.

That same year I moved fish, tanks, and toxics from Long Island and set up my laboratory in a small brick outbuilding all but abandoned by the University of Massachusetts. My husband's lab offered me nothing in the way of a job or even a potential job. But that year, as the USGS biological laboratories struggled for their lives in a climate of Republican antienvironmentalism, they were not in a position, nor was the laboratory director inclined, to consider hiring "the wife." So on a blazing hot day that summer, with Sam on my back, my white shirt stained with sweat and creamed sweet potato, I met with a benevolent department chair at UMass who offered to process my application for adjunct faculty status. This would allow me to apply for grants (albeit as a co-principal investigator only). In deference to his own faculty, the chair warned me not to "crowd" the newest hire, an aquatic toxicologist like me. Other than writing grants, I had no sensible long-term plan except to apply for that aquatic toxicologist's position if and when he either transferred to a different university or was refused tenure. So a year later, when Sophie, our second child, was born, I was for the first time without a grant and without a plan.

A couple of months after Sophie's birth, with manuscripts completed, grant reports submitted, and Sophie at my breast, science was far from my mind when the phone rang one day. Phrases like "polychlorinated biphenyls (PCBs)," "we need an expert," "striped bass," and "Hudson River" emanated from the receiver. Sophie squirmed and milk dribbled down my nightshirt as I struggled to find my inner scientist. Out of the blue, it seemed, I was being asked to consult. Once again, I was off and running, this time on a path I hadn't considered. No fieldwork, no grants to write, no nasty chemicals to inject—just data review and synthesis, and all from my upstairs office. Two years later, as the project came to a close, a contaminant specialist position opened at the regional U.S. Fish and Wildlife Service. It was, I had thought, just the kind of work I'd love to do—similar to the consulting job but permanent. "We could consider moving the position closer to you," said the supervisor when I balked at the hour commute. That would certainly help facilitate the fifty-hour workweek he'd described. But by that time I'd worked on my own, project by project, for ten years. The kids were three and five years old. I was independent and had worked part-time ever since they were born. The thought of both at once—losing my independence and full-time work—did not appeal, and I declined the offer.

In fact, I had become so committed to being home for the kids that when the UMass toxicologist position finally did become available and a friend in the department urged me to apply, I submitted my application with a cover letter indicating my commitment to part-time work, should they be interested in considering a shared position. Though I'd been advised over the years not to negotiate until *after* the job offer, I found it difficult to be dishonest about a situation I knew I couldn't accept. That was the last I heard from UMass.

Though my request for part-time work didn't impress the selection committee, neither, I'm sure, did my more than ten years of part-time research, consulting, and hodgepodge of a career. I had worked myself out of one of the most traditional choices for a scientist with my experience: research and academia. I realized for the first time that I had also left behind, possibly for good, the joy of laboratory research—completing an experiment, making sense of a unique data set, and the thrill of discovery. Occasionally I mourn my passing as a research scientist, the one whom I had envisioned as a girl. Then I come to my senses and appreciate the freedom that I have had since leaving the lab to explore the field of environmental toxicology—experiences I likely would not have had in a more traditional career track. Over the years I've met and worked with a large cast of characters, and for the most part, I've looked forward to each new adventure. I have had the opportunity to work with and learn from maverick scientists who are leaders in their fields and from citizen activists worried about depleted uranium in their neighborhoods, PCBs in their rivers, solvents in their groundwater, and the ubiquitous military hazardous waste sites. I have worked with Gloucester fishermen interested in the effect of contaminants on cod stocks and with students eager to help communities affected by contaminants in any way they can. I've published in respectable scientific journals as well as in local papers and magazines introducing the lay public to my passion, toxicology. I owe my freedom to an Internet full of scientific resources and university databases and to the financial and emotional support provided by Ben, who understands that working off the scientific grid is difficult. Each time I've seriously questioned my choices to go it alone, he has offered to rearrange his work and time, to essentially trade places, should I desire to redirect my career.

Despite all the advantages of my independent way of life, there are many times when I have questioned my choices, sometimes feeling a twinge of guilt over the federal money that has supported my development as a scientist, which ironically has also granted me the privilege of choosing how I work even if it does not fulfill expectations of funding agencies and academic advisors. There are the times when I must pull my head out of whatever it is I'm working on to pick the kids up at school or take them to the dentist or to music lessons, losing a train of thought that took hours to develop. There are also times when I resent that I am the one who volunteered to be the housewife and when I peruse the job announcements (each requiring the applicant to be on location in Washington, DC, New York City, or Boston) and wonder if it was wise to settle out here, hours from the major cities.

But the worst times are when the guilt and regret that I am neither the mother I imagined I would be nor the researcher that I'd envisioned as a graduate student spill over into the family. That is when I snap at the kids and when I wonder if the trade-offs that I have made to stay home, working part-time, were worth it. Although I cherish afternoons with my children, sitting by the river watching as Sophie dips her big toe and then her whole body into the rushing water, and as Sam cannonballs from a fallen tree, my afternoons at home are not always so idyllic: Get off the computer, do your homework, did you practice? go outside and play, *Sam please—get off that computer.* So when I recently asked Sam (now thirteen and more responsible) how he felt about my being home, knowing that several of his friends came home from school to empty houses—computer and video game heaven for the neighborhood boys—Sam barely had to think. "Comfortable," he said. "I felt comfortable and kind of safe, knowing you were around." I knew exactly what he meant, having felt the same comfort returning home when I was a child. But must my basic desire to provide comfort, to be available for my kids, clash with my other lifelong love of science? As I am now realizing how much of my professional self I've traded and altered to achieve a work-life balance, I can't help but wonder how my career might have evolved had there been more options. Part-time faculty or federal positions? A spouse friendly employer? More resources for independent scientists? Although I've pieced together a rewarding career that accommodates a life with my family that feels right, I hope the next generation will have a wider array of opportunities, and that the federal, educational, and private institutions that court the young woman scientist will continue to support her as she matures as a scientist—and if she chooses, as a mother.

Geological Consulting and Kids

An Unpredictable Balancing Act?

Debra Hanneman

Geologist, Whitehall Geogroup, Inc.

PhD, University of Montana, 1989

Nineteen eighty-nine was a banner year for me. I finished my PhD in geology in May and then had my first son in December—not exactly the easiest path to career advancement. But with what was going on in the rest of the world with the Berlin Wall falling, the Tiananmen Square uprising, and the U.S. invasion of Panama—I couldn't complain too much. I was actively looking for a position in academia, but I was also doing consulting work in the earth sciences. As a few more months went by, my preference for consulting work began to take precedence over any thoughts of getting a position somewhere in academia. I found that I could decide when to work—at least this happened most of the time, when I wasn't taking care of my son, Ben. I also took him with me to the field. By the time Ben was six months old and ready to ride in a backpack, the weather in Montana was good enough for both of us to be outside doing geological fieldwork.

I decided that the life of a consulting geologist would be the way to go —at least for a while. Consequently, I incorporated my own company, Whitehall Geogroup, Inc. And within that same year of starting my company, I had my second son, Matt. Having two small boys did complicate doing fieldwork. Luckily, I have a really supportive husband who loves being with his sons. Moreover, he is a geophysicist who likes to get out of the lab and look at rocks. So we would all pack up in our four-wheel drive vehicle and go to the field.

Do not conclude that all our field excursions were wonderful family times together. We went through all the scenarios of crankiness, getting sick, and just pure boredom that all families who travel together experience. We also had the added excitement of things that happen to fieldworkers. On one trip, for instance, one of my clients came madly running after us to tell us to keep our boys close at hand because a mountain lion had been prowling around just a few hours before we started work on the property. Then there is our fond memory of doing fieldwork one winter day near Yellowstone National Park. My husband went off in one direction to map out an area, and I drove off in another direction with both boys and a ranch manager. Within a few miles of leaving the paved road, I got our vehicle stuck in a rather large snowdrift. While the ranch manager hiked four miles back to get a tractor, Ben and I dug us out of the drift. Luckily for Matt, he was still in his car seat. I am constantly reminded of that episode whenever the winter snows hit and I am driving.

Several years have now passed since I started both my family and my company. Many changes have taken place in both and in how those two areas of my life interact. Throughout this time, I have assumed that I work the equivalent of a full-time professional position, but, as with any self-employed person, sometimes it is difficult to determine exactly how many hours of billable time per day one works. Basically, I focus on project completion rather than on hours worked per day. Consequently, some workdays are much longer than others.

The one constant with my yearly work schedule is that I always set aside time—particularly in the winter months—to compile my research. In doing so, I've been able to continue to publish professional papers and abstracts. There are a few time gaps in my professional paper publications that correlate to increased business workloads and family events. Meeting abstracts are, of course, much less time-consuming, and therefore I've been much more prolific with them. The bonus with meeting abstracts is that I also attend a professional meeting, so there is some added incentive for abstract writing.

In regard to changes in my life and work over the past years, let me begin with changes in my company. When I first began consulting, I did a large amount of work on mineral remoteness tests. These tests are an important component of conservation easements and are ultimately used by the IRS for tax purposes. Thus my contracts were largely with land trust and conservation organizations. Other types of projects that my company picked up were in the areas of hydrology, hydrocarbon exploration, and geologic mapping. These were all private entities that my company contracted with, and hence I put together the contractual agreements. I had a very good template to use for a general contract, and I also asked two attorneys to review it so that it covered my company sufficiently. I still use this same contract and have added only a disclosure clause to it, so supposedly I know where and how my work is being distributed.

Throughout the early years of my company, my boys went from preschool to public elementary school. Because I set up my own contracts, there was no problem with policies regarding my sons' accompanying me during my work, so it was fairly easy to take them with me when I did fieldwork. And I could get office work done with them at home because I worked in a home office. My hours of office work, however, were not usually the nine-to-five type. My time frame became the project deadline as specified in my contract, and I molded my work schedule to accommodate it.

Since the late 1990s the types of projects that I take on for my company have changed, and my family's involvement in my work schedule has also undergone modifications. The dot-com bubble burst of the early 2000s negatively affected the large-acreage Montana real estate market and thus slowed the conservation easement business. The situation for my company's work, though, was not as dire as it could have been. I had already come to the decision that several of the easements I worked on were simply tax write-offs. By this time, I had also decided which organizations I preferred to work with, and I have continued to work with them.

In the meantime, wireless Internet had come to rural Montana. That opened up numerous opportunities for doing online projects from my home office. All I needed to do was to invest some time and money into incorporating high-speed Internet service into my office, and I was open to other kinds of project work. This brings up a point that needs to be made about running a home office. It is not difficult to set one up, but the office equipment needs to be kept running, updated, and finally disposed of. Depending on what is going on, particularly in the cyber world, this may take large, unplanned segments of time out of a scheduled workday. Another problem I often encounter is the assumption that if you are home, you are not working. But, for any multitasking professional worker, these are not insurmountable problems.

The gradual shift in my work from largely fieldwork-based projects to online work coincided with my sons' going through middle school and on into high school. No longer did I need to take them with me during field-work, and they usually did not want to go along anyway. Add to that the ever-greater commitment of after-school activities—particularly in high school—and I find myself setting my work schedule to accommodate day-time and weekend school events. In some ways it was easier, at least for time scheduling, to work when they were younger. The primary difference I find is that it is far easier to concentrate on work now that they are in-volved in their school activities and, more important, can drive themselves. Because of an improved concentration level, I find that I can complete projects more quickly. Overall, although it seems that I spend more time now with my boys at their school functions than when they were younger, I think that my work schedule goes much more smoothly because I can once more concentrate on work for longer, uninterrupted time intervals.

The change from mainly fieldwork-based projects to online or at least largely digital-type projects has again caught the interest of my sons. Some evenings find us all in the home office area. They are not really there to see what I'm working on, but are more likely involved in doing their home-work, playing iTunes, or instant messaging to their friends. Occasionally, when I'm updating the company website, putting the finishing touches on a CD presentation, or putting together an online quiz, I'll show them what I'm trying to do. We've had some good laughs, and I've received some re-markably honest input from them on the merits of my projects. Sometimes my sons' friends also drop by and of course are drawn to the home office area. In the end, I have an unlimited supply of feedback on my work, al-though it is primarily from teenage sources.

All in all, my decision to work as a consulting geologist has been a good career path for me. There have been frustrations—there is no way to get around the fact that raising a family is time-consuming and mentally drain-ing even with the best of help. But I have been able to continue working as a professional geologist and have published several professional papers since my sons were born and since I started my company. And, to my great satisfaction, my family and I are still growing in many different directions.

Career Scientists and the Shared Academic Position

Carol B. de Wet

Professor, Department of Earth and Environment, Franklin & Marshall College

PhD, Geology, University of Cambridge, England, 1989

For me, it all began twenty-six years ago, in 1981, when I was a senior in college and attended a geology lecture. I was an undergraduate at Smith College and had grown up in an environment where gender was never considered a reason for not doing something (my father taught astronomy at Wellesley College). But on this particular occasion the guest lecturer, in responding to a question after her talk, made an assertion that changed my life. She said it was "impossible for a woman to have a successful career in science and be a mother." I don't remember whether there was silence in the room immediately after her response, but there certainly was for me. Could this be true? Why would this be true? And could I prove her wrong?

For the past seventeen years I have shared an academic job with my husband. We received our PhDs from the University of Cambridge, England, married in England, and moved to the United States. I received a job offer from Franklin & Marshall College in Lancaster, Pennsylvania, and requested that the college consider hiring my husband as well, in a shared position.[1] This meant that the college would get two faculty members for

---

1. C. B. de Wet and A. P. de Wet, "Making It Work Together: Spouses on the Tenure Track," *Geotimes* (April 1995): 17–19.

the price of one but that we would share the teaching load.[2] To the administration's credit—this was back in 1989 and fairly novel at the time—they agreed. My husband and I worked independently for tenure, successfully received tenure, and have rewarding teaching and research careers.[3] During the same time period we had two more children (our first child had been born during my postdoctoral work). The oldest child is about to leave for college next fall. So, looking back, can I now say to myself that I proved that woman geologist wrong?

The answer to that depends on how you define success. If your definition of a successful geoscience career in the academic world means—as I once heard a woman geoscientist describe it—that you are running multiple National Science Foundation grants simultaneously, have a cadre of graduate students, and are named to national scientific panels, then my career might not fit into the successful category, and the speaker might have been right. If, however, your definition of success includes satisfaction through teaching, publication of research that has significance for you, some time for yourself and your family, and feeling that you gave your children the strongest foundation for life that you possibly could, then that speaker's supposition might be incorrect. Ultimately it comes down to a personal view of one's life. There is no one correct definition of success, nor is there a single pathway to achieve it. The shared job was the right path for me in that it provided a mechanism for my career to continue to move forward in research and scholarship while giving me the gift of time to be with my children as they were growing up. Now, as I am poised to see my oldest leave for college, I am struck with how short the period of intense family time really is, relative to my working life. I will still have nearly fifteen years to devote to my geoscience career after my youngest child leaves the house for college!

I do not see the shared position as part of a negative mommy track. I have had the same career opportunities as my full-time colleagues. I have applied for, and received, a National Science Foundation grant and Petroleum Research Fund (American Chemical Society) grants, and I have pub-

2. C. B. de Wet and A. P. de Wet, "Sharing Academic Careers: An Alternative for Pretenure and Young Family Dual-Career Faculty Couples," *Journal of Women and Minorities in Science and Engineering* 3 (1997): 203–12.

3. C. B. de Wet, G. M. Ashley, and D. P. Kegel, "Biological Clocks and Tenure Timetables: Restructuring the Academic Timeline," *Geological Society of America Today* (November 2002): S1–S7.

lished research, generally coauthored with my research students, at a rate comparable to my many of my peers at other liberal arts colleges. I was an officer in the Geological Society of America's Sedimentary Geology Division. I have had the opportunity to work closely with the administration on family-related policies that affect the college's faculty and staff. I have not felt any denigration from professional colleagues about being technically half-time, although they may simply be too polite to say something directly to me. Most of the women I have spoken with ask about the shared position with a sense of curiosity and interest. In our professional lives we constantly make choices about our priorities, whether to put our names forward to serve on professional committees, whether to write that next grant proposal, whether to attend every professional meeting, whether to do fieldwork in a remote location or closer to home. Since the shared position does not preclude any of those choices, the only significant drawback I have experienced is the single salary. That really amounts to a lifestyle choice. For us it meant that we didn't eat out as often as we might like and that I taught my children how to help clean house and do the laundry rather than hiring a cleaning person. But ultimately, individual families always have to define their priorities and accommodate them according to their values and finances, regardless of their salary level.

What are the practical implications of shared academic positions? Most institutions expect that you and your spouse (or partner) will be in the same department. It is difficult (although not impossible) to ask a department that has requested a full-time position to manage with only half as much teaching as it expected, while another department gets the "gift" of a half-time person. So it is simpler if your fields are closely related and you both are hired in the same department. It is, however, better for the department if your subfields are different enough that having two of you can significantly broaden the teaching and research scope of the department. This is particularly relevant in small departments, where one more person's expertise can have a large impact on the curricular offerings. Being able to articulate how hiring the two of you will give the department double the expertise and double the breadth of teaching is a powerful argument in your favor.

As far as I know, it is important that each person be evaluated independently for tenure and promotion. Although it would be difficult if one person got tenure and the other did not, it would be unfair to jeopardize both careers if it turns out that one person in the shared job is not well suited to academia while the other person is. Careers may move at different paces, and separate evaluations allow for greater flexibility for both the couple and the institution. The fairest contract language should state that tenure and promotion should require "half the quantity but all of the quality" expected of any full-time faculty member. I think this should apply to both the teaching load and the publication expectation. The shared position was not conceived of as a way to shirk teaching in favor of more research. At least for us it was envisioned as a way to manage two careers and a family. Therefore, imposing a full-time research expectation onto each member of the shared position implies that the family time component is not valued or important.

In terms of other nuts and bolts, we each have one-half vote in department meetings but a full vote each in college faculty meetings. We each have an office, a laboratory, and full access to all college professional development funds and opportunities. In exchange, we end up supervising more, and more diverse, student research projects than a single faculty member could. We take more students on field trips and offer a wider range of courses than a single faculty member could easily accomplish.

Shared positions do not work well if the faculty members approach them as a way to sneak their partner into a job or if they see the shared position as a way to negotiate their way to two full-time positions. Conflict may result if the two faculty members demand to fill every departmental vacancy caused by sabbaticals and leaves or expect to temporarily become full-time employees. Both the faculty members and the administration need to be very clear during the hiring process about what the work expectations and limitations will be. We have found it most helpful to be flexible wherever possible: for example, if in one year both of us need to be on time-consuming committees, the next year the provost will suggest that our committee load be lessened. This works similarly for advising and even for our teaching load. If, for some reason, the department needs us both to take on a slightly heavier load for one or two years, the expectation is that the load will be lighter the following year or two. It is really just a process of sound record keeping and maintaining goodwill.

When my husband and I proposed job sharing, we thought of it as a practical solution to the two-professional-body problem. We both wanted academic jobs and we wanted to be in the same location. The shared position provided us with that. I did not have the hindsight I have now to see the job-sharing option as a way to balance family and career demands over the approximately twenty years it takes to raise three children. From my perspective the speaker I heard all those years ago was wrong. Each scientist will have to make that decision for herself, but fortunately there are a number of career options, the shared academic position being one of them, to help guide her decision.

SECTION III

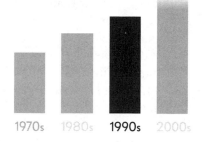

1970s   1980s   **1990s**   2000s

MAKING ROOM FOR BABY

"Ladies," said the policeman, "I wouldn't go there if I were you—and you, especially in *your* condition." He looked down at my pregnant belly. Adria and I exchanged glances, thanked him, and headed on toward the Roanoke Yacht Club, a small cinder-block building somewhere behind the loading docks at Newark Airport. The "club" was one of our contaminated field sites; but it was also a gathering spot for some of the locals and a place where one would more likely expect to see bodies floating under the bridge than yachts.

Several weeks later, as our first field season neared its end, I hauled two buckets of killifish from Flax Pond, a nearby reference site, up to the laboratory. The moon was full, and the fish were gravid. That night, the moon's pull influenced more than just the tides and the killifish. The waters protecting and cushioning my own baby broke, a torrent of warmth gushing down my legs. Sam was born the very next day, June 25, 1994; and my life as a scientist was forever changed.

. . .

In the early 1990s two issues of *Science* were devoted to women and women's issues: twenty-four of their valuable pages in 1992 and forty-nine in 1993. Daniel Koshland Jr., then editor of the journal, noted, "As society searches for solutions to the horrendous global problems in need of scientific input, we cannot afford to lose the potential of women's brainpower." He added, "In simple fairness, the playing field must be leveled so that women are not inhibited by a less than helpful environment."[1] In 1990s the Association for Women in Science celebrated its twenty-fifth year, and Rita Colwell, a microbiologist and oceanographer, became the first woman to head the National Science Foundation. In 1996, although MIT reported over 100 women on its faculty, out of 209 tenured faculty in the School of Science, only 15 were women. A committee established to study the status of women faculty at MIT reported that

> junior women faculty feel well supported within their departments and most do not believe that gender bias will impact their careers. Junior women faculty believe, however, that family-work conflicts may impact their careers differently from those of their male colleagues. In contrast to junior women, many tenured women faculty feel marginalized and excluded from a significant role in their departments.[2]

For women in science, the differential impact of family on their work lives in the 1990s reached beyond academia. A report issued by the National Academy of Science found that for women with doctorates in science and engineering, "being married and having children were associated with relatively high unemployment rates; for men they were associated with relatively low unemployment rates."[3] Another study, this one of workers in the Biotechnology Industry, also found differences in the potential impact that family status had on scientists, reporting that while men with children tended to have partners who worked part-time, the partners of women with children worked full-time, and those with greater support at home tended to be promoted "furthest and fastest."[4]

1. Daniel Koshland Jr., Editorial, *Science* 260 (1993): 275.

2. The MIT Faculty Newsletter, "A Study on the Status of Women Faculty in Science at MIT" (March 1999), http://web.mit.edu/fnl/women/women.html.

3. National Science Foundation, Division of Science Resources Studies, *Who Is Unemployed? Factors Affecting Unemployment among Individuals with Doctoral Degrees in Science and Engineering* (Arlington, VA: National Science Foundation, 1997).

4. Susan Eaton and Lotte Bailyn, "Work and Life Strategies of Professionals in Biotechnology Firms," *ANNALS, AAPSS* 562 (1999): 159–73.

Despite these inequities, by the mid-1990s women with either master's degrees or PhDs in science or engineering represented almost a quarter of those employed at four-year colleges, and an even larger proportion of women worked at private nonprofits (42 percent) or were self-employed (32 percent.)[5] In 1999 the NSF reported an increasing percentage of women in the faculty ranks, with women representing 37 percent of junior faculty, reflecting the increasing share of PhDs earned by women. Moreover, noted the NSF, the positive trend should continue, assuming that "women stay in academic positions at a rate equal to or greater than men."[6]

5. National Science Board, "Science & Engineering Indicators—1998" (Washington, DC: National Science Foundation, 1998), table 3-12, http://www.nsf.gov/statistics/seind98/access/append/appa.htm.
6. National Science Board, "Science & Engineering Indicators—2002" (Washington, DC: National Science Foundation, 2002), chap. 5, http://www.nsf.gov/statistics/seind02/c5/c5s2.htm.

Less Pay, a Little Less Work

Heidi Newberg

Associate Professor of Physics, Rensselaer Polytechnic Institute

PhD, Physics, University of California, Berkeley, 1992

I have walked the straight and narrow path to success in physics. I went from college to graduate school, from graduate school to a postdoc, accepted my first tenure-track position five years after my PhD, and was tenured at age thirty-eight. At no point in my career did I ever dare to dream that this would be possible. And even now, I wonder how long my career in research will last and how successful I can continue to be. I never doubted my own intelligence or skillfulness. But I have always known I wanted children and a happy family more than I wanted a successful career as a physicist/astronomer. I will not dedicate my whole life to science as some around me do. And that is where my career ambitions have wavered. Is it possible to do both? Is it possible to be a successful scientist, to make discoveries, without dedicating one's full attention to career ambitions?

I came up against this question when I was a senior in college. I remember meeting a Nobel laureate while visiting graduate schools. I told him I really liked hobbies like music, drawing, crocheting, and dance, and I wanted to know if I could still keep up these hobbies and go to graduate school in physics. He told me that learning physics was really, really difficult, and it probably was not worth doing unless you dedicated yourself to it full-time. One has time for hobbies, he said, after retirement.

This well-meaning professor answered the way I expected him to, though not the way I had hoped. I had worked hard all my life—first on our family farm and then in college. I was not afraid of hard work. But really, dedicating my life to science at age twenty-one? Five to seven years of graduate school seemed possibly more than I was willing to commit to. And having babies in retirement was an unworkable plan. But when I received my acceptance letter to go to the Berkeley physics program, a feeling in the pit of my stomach told me I wanted to go. I have always taken things one step at a time, asking myself at each step if I am still using my life the way I want to, yet wondering at all times whether the system will deflect me from my goal.

I have made decisions in my scientific career with the aim of balancing my career with family life. I took a postdoc position at a national laboratory because it was likely to last five years, and this would give me time to start a family. I was also offered a postdoc position at Harvard, but though it might have been more prestigious, it was only a three-year position and was guaranteed for only one year at a time, subject to incremental funding. Unfortunately, I had difficulty conceiving and did not become a mother until the last year of my five-year postdoc. I stayed at the national laboratory for a total of seven years, and though I expect that conditions there were much better than for postdocs at research universities, there are ways in which they could have been improved. Some things, such as day care close to office areas and facilities for using a breast pump are simple, but the primary need of a new mother is time. She needs time to take care of her child and herself.

We adopted our first daughter, and one year later I gave birth to a second. My children came at about the time I was up for promotion to a tenure-track position at the laboratory. I succeeded and received the tenure-track position but only after a fight. I took three months of leave for each child, as did my husband, who was a member of the educational staff at a nearby research university. Juggling our schedules, we each worked three days per week and took two days of leave so that we could keep the babies at home as long as possible and still maintain our projects at work. My employer required that all of my vacation time be used up as part of the guaranteed three months of (otherwise unpaid) leave afforded by the Family and Medical Leave Act. My husband negotiated a less formal arrangement in which he worked 3/5 time and did not automatically lose his vacation, so he was able to extend his leave somewhat by using vacation days. When my children were one and two years of age, they were both in day care full-time. At least one of the children was sick every other week. Before I became a parent, five weeks per year of vacation time seemed infinite. Now my vacation coupled with my husband's seemed barely adequate to cover the days when the kids were sick. By the time my older daughter was almost three and mostly potty-trained, I had just begun to build up a cushion of vacation time so that I wouldn't have to worry when the kids were sick. Maybe we could even think about taking a real vacation. Then, at a kids' birthday party barbecue, my younger daughter put both hands directly on the charcoal grill and received second-degree burns on both palms. My nerves were shot. My daughter was home from day care for a week with bandaged hands and required almost constant care. My cushion of vacation was gone.

I returned to work, but I was beginning to feel that I could not keep going this way. I asked my supervisor if I could work 80 percent time. I planned to work four days per week—whichever four days neither of my kids was sick. He agreed, but the division head would not sign the papers without talking to me first. The division head thought I might be able to work this out, without losing pay, by working part of the time from home. I tried to explain to him that one- and two-year-olds don't play by themselves very long and that they need a parent to care for them when they are sick. He told me that he and his wife didn't have any children, and they just worked all the time, a comment to which I wasn't sure how to respond. To his credit, he signed the papers though I don't think he ever understood the problem, and I worked 80 percent time for one year. In reality, I'm sure I often worked forty hours per week, which is more than I was required to work but less than postdocs are typically expected to work, and I didn't feel guilty about the time I spent with my children. It amounted to about two months of leave and may easily have saved my sanity and my career.

When my children were two and three years old, I was offered a job at a research university located near my extended family. The opportunity came at a time when my relationship with the national laboratory was strained, and though I had some concerns about juggling a family and a faculty position, I took the lateral move. I wouldn't say taking on a faculty position is easy—with or without children—but I worked through learning how to teach, advise, and write grant proposals using extensive time-management techniques and constant prioritization. I was awarded tenure after five years on the faculty.

My husband and I have recently added two more children to our family, one adopted and one by birth, so I have now experienced new motherhood at a university. On paper, my university has a great family leave policy: one semester without teaching (nominally a two-day-per-week commitment) and an optional additional semester without teaching at half pay. The trouble with the policy is that decisions on how to implement and pay for this benefit are left to individual department chairs, deans, and the provost, sometimes with unsatisfactory results.

Fortunately, for me the policy has worked exactly as intended. My younger children arrived sixteen months apart, so I have taught one semester in the last five semesters. I have been able to keep up my research, write grants, advise students, and fulfill my service functions to the university while spending some extra needed time with my children. Because I need to have child care covered for after school, school holidays, days when any of my four children are sick, and full-time care for two very young children, we have employed a nanny this time around. It is possible to hire wonderful young women to care for children if one is able to pay a living wage, which is rare in the child care industry. Our child care bills alone, which are mostly paid after taxes, eat up half to three-quarters of my (reduced) income. We are lucky to be able to spend at a small deficit for a few years, before preschool and public schools significantly reduce our child care needs. I remember thinking as an undergraduate that a faculty position would be the perfect job for a mother because the hours are flexible and it is easy to reduce the workload without changing the job description—one just teaches fewer classes. In some ways it has been easier, because the university really can lower the workload by reducing the number of classes I teach.

The part that does not work is the same part that did not work at the national lab—the perception of my supervisor and my colleagues is that I am not working as hard as everyone else. I have tried to educate people that I am paid less to do less work. I have tried to educate my peers that I am not on sabbatical; I am still on campus doing everything I would normally do except teach. And when I am on half pay, I work much more than half-time. Currently I work half-time and pay a forty-hour-per-week nanny to take care of my children. Part of what fuels resentment is that other faculty members are asked to teach the classes I normally would teach instead of or in addition to their normal class loads. Although the university handbook for faculty states that funds will be provided to cover a replacement instructor during my family leave, there is no money actually set aside anywhere in the university to cover these costs: the money must be squeezed out of the budgets that are already in place.

My children are now aged eleven, nine, two, and under one—two sets of two. Now I am again feeling the strain in my relationship with my employer. Although I recently survived the tenure process and have garnered international attention for my research, administrators at the university have told me I am underproductive. This attitude reminds me of the discussions I had with administrators at the national laboratory when my older two children were toddlers. Though it is impossible to know for sure whether there is a causal relationship between my employers' stated negative perceptions of me and my motherhood, the similarity in timing—coinciding with the birth of young children—makes me wonder if the real problem is my decision to take time off to care for children.

Years ago, a forty-year-old friend of mine who had resigned her position in the management of a bank after adopting two children told me that they look at you differently when you work part-time. The kinds of assignments and the available options shrink. "You'll see," she told me. Now I am forty, and I think I see. I am no less able or ambitious, and I am no drag on the organization that pays my salary. When I did less work, I took less salary in accordance with the written and agreed employment policies. I do not expect to be mommy-tracked now.

As I do often in my career, I am again asking myself if I am doing what I really want to do. A few months ago, I asked my nine-year-old daughter if she cared whether her mother remained an astronomer. She got a long look on her face and told me she liked my being an astronomer because I could help her with things that other mothers couldn't. She smiled when I assured her that even if I weren't working, I would still know everything I know now. When I was a young girl, a friend of my parents who was struggling to win the respect of her academically gifted son told me that education is never wasted on women; I guess she was right. And boy, do I have a lot of education!

Last month, I was awarded two National Science Foundation grants back-to-back. I gave an invited talk at an international conference, and I was invited to visit China to give several lectures, all expenses paid. I am not ready to leave science, but even as a forty-year-old tenured professor, I'm still not ready to commit my life to my job.

Reflections of a Female Scientist with Outside Interests

Christine Seroogy

Assistant Professor, Department of Pediatrics, University of Wisconsin

MD, University of Minnesota, 1993

Gender bias in science clearly exists. One has to wonder if this is because for some women, work is secondary to family responsibilities or whether women lack the aptitude to be successful in the pursuit of a scientific career. The data suggest the former conclusion.[1] For example, women who have children soon after receiving their graduate degrees are much less likely to achieve tenure than their male counterparts.[2] Recently a prominent female scientist at a renowned institution on the East Coast shared with me her own experiences with female graduate and postdoctoral students who had children during their time in her lab. She found it difficult to predict a priori which female scientists would continue along the path toward full-time academic positions and which ones would choose alternative career paths. In her experience, the outcome was evenly split. I found this shocking but then reflected on my own training experience and remembered the many pregnant colleagues with whom I have interacted. The fifty-fifty split was consistent with my own observations. The impor-

1. Survey of Doctorate Recipients. National Science Foundation, Division of Science Resources Statistics, 1993–2003. http://www.nsf.gov/statistics/.

2. M. A. Mason and M. Goulden, "Do Babies Matter (Part II)? Closing the Baby Gap," *Academe* (November–December 2004): 3–7.

tant question becomes, do these alternate career paths develop by choice, or are they forced upon women by the current environment for female scientists in academia? In my own career, I almost avoided the bias (or "maternal wall") entirely by deferring or simply not having children, but thankfully I committed to this important life-altering experience twice. My mantra now for young women in science is that there is never a "good" time to have children, but doing so is well worth the accompanying challenges of balancing career and family life. As I search for role models on how to advance from assistant professor to a tenured position, I have become dismayed. Having observed the career path of the highly successful female scientists at the numerous institutions where I have trained or worked, I can make some generalizations about the gender differences at this level. Many successful male scientists have a multitasking primary caregiver wife who tends to carry the domestic load, or at the other side of the spectrum, there is the male scientist whose partner or spouse is an academician and typically a few steps behind him on the career path. In contrast, the highly successful female scientists advance their careers within a very small spectrum. Either their partner/spouse works full-time outside the home and typically holds a high-powered job or they are single. Sometimes, when I recall certain former female colleagues, I shudder a bit and tell myself, "I just don't want to be like her." I don't want to be perceived as someone who is rigid, overly assertive, and not available for mentorship or unapproachable. This is the catch-22 for high-achieving female scientists. It is not to say that, as a result, I seek male mentors exclusively. I have received great advice from male mentors, but I desperately seek the female perspective and find it difficult to obtain.

One of the challenges of balancing a family life with an academic scientific career is that the pursuit of answers always generates more questions. Being adequately prepared to ask important questions, interpret the data, and formulate the next question is a consuming yet thrilling endeavor. This is coupled with the necessity of writing manuscripts and grants to ensure success. Add raising children to one's responsibilities, and the ability to maintain intellectual stamina for a scientific career can sometimes feel hampered. On some days it can seem overwhelming, and at the end of the day I often ask myself what I really accomplished. My role as a physician-scientist adds patient care responsibilities and another skill set to maintain. A mentor recently described for me a time-management style that promotes efficiency. Tasks are ranked according to importance and urgency. On its face this prioritization schema seems logical and effective. For the physician-scientist patient care trumps everything else. It is always urgent and important. For the mother who is also a physician-scientist this poses a dilemma because important children's activities, such as their first school play, will be eclipsed by an urgent patient-related matter.

During my training I had the good fortune of working alongside larger-than-life physician-scientists of both genders (though not surprisingly, more men than women), and I distinctly remember one conversation with the chair of my department, a highly successful and preeminent male physician-scientist. I was shocked when he changed the course of our medicine-based conversation and asked about my relationship with my father and, more specifically, about my father's availability during my childhood. I told him our relationship was very good: he was a high school biology teacher and was always home for dinner. He coached some of my sports teams and was almost always present at important events in my life. These were my recollections. My department chair then confided that one of his regrets as a father was that he was not around enough for his children while they were growing up. I have heard similar sentiments from other male colleagues in senior academic positions, but I have yet to meet a woman in an equivalent position who has expressed the same regret. Why? I believe that women, in general, do not permit themselves to speak these thoughts out loud because they are too painful. Their struggle for success and tenure requires tremendous sacrifice and conscious decision making against mainstream expectations, whereas for men, the path toward becoming a scientist and establishing a family is clear and obvious. Women with full-time careers must navigate a path for balancing work and family that is counter to societal traditions. Personally, I struggle with the ability to be available for important moments in my children's life and maintain my professional responsibilities.

On a recent grant submission I was required to contribute a personal statement detailing my career goals. The majority of the statement was a review of my career up to this point, demonstrating my strong commitment to an academic career. I talked about the importance of having a "fire in the belly," meaning a persistent temperament that can deal with the challenges of science and unrelenting curiosity. I wrote that my long-term career goal was to achieve tenure at an academic institution. At the close of the statement I added the following:

> The other aspect that I strive to model for the many students with whom I interact is that life is about balance, and that a scientist who spends all of his

or her time consumed with his or her work (and yes, it must be a passion or one is doomed to fail) and lacks outside interests is creating a situation that will truly hamper the intellectual creativity that is necessary to be an outstanding scientist.

I realize that "outside interests" can mean many things to many people and that it may seem callous to label family an outside interest, but I guess that is as bold as I ventured to be. The problem is that in the world of science, family *is* considered an outside interest. How can the two worlds—pursuing a full-time academic career and raising a family—coexist? Many discussions need to occur to improve this situation. The current movement toward balancing faculty careers and family life is gaining momentum to collectively improve the work environment. The climate has changed during my short time in academia, and I believe it is a necessary change—not just for female scientists but for male scientists as well. For example, tenure-clock extensions are being provided for women (and men) who have children during this critical period in their careers, demonstrating an increased sensitivity to scientists' desires to balance work and family. Some of the changes that I have witnessed are the beginning of open dialogues that address the challenges of combining academic scientific careers with raising children and recognize the need to improve the work-life balance for academic faculty. We can do better, and I remain optimistic that collectively this dialogue will translate into better policies for both genders.

Part-Time at a National Laboratory

A Split Life

Rebecca A. Efroymson

Environmental Scientist, a U.S. government research laboratory

PhD, Environmental Toxicology, Cornell University, 1993

The most challenging day of my split life between motherhood and science is a blur. The manicurist at National Airport watched us make a dozen laps around her cart, sometimes clearing it, sometimes crashing into it, and sometimes entangling ourselves in its legs. I used to wonder how many women had layovers long enough for an airport manicure (and nails long enough to want one), but that thought was the furthest from my mind as I attempted to corral my toddler between laps and temper tantrums. I was thankful for three things—that I had checked our luggage, that I had not needed a laptop, and that I had brought Sam's harness. A colleague had recommended the harness as she fondly recalled her son circling her legs next to her poster at a conference more than a decade earlier. Whereas her son's behavior had been like a well-behaved pet, my boy's was more like a pack of wild dogs.

Earlier that day I had left Sam, feverish and screaming, in a Washington, DC hotel room with my parents, whom I had brought to town from Philadelphia via Amtrak. I had traveled to DC to defend a grant proposal to the Science Advisory Board (SAB) of the environmental research arm of a government agency that was supposed to cover a substantial fraction of my salary and that of a colleague at the federal research laboratory where we work. Unfortunately for my parents, the only calming technology they had in their arsenal was the ice machine down the hall. My husband was traveling as part of a prior commitment to a National Science Foundation committee, so Sam had flown to Washington with me. The SAB defense was the last step following peer review to secure funding. Though I believe that members of the SAB were critical of my proposal even before my presentation, my lack of sleep certainly contributed to my failure to interpret and answer their questions adequately that day. Moreover, I was distracted by the sliding schedule of the presentations in relation to the hotel's grace period for late checkout. Even had Sam not been feverish, it was too cold for my family to wander the streets waiting for me. This was the first time that my split life might really have impacted my work and the viability of my job.

I took this project loss hard. I thought I had a great idea for establishing relative causality for various threats to rare wildlife species on military installations. My chagrin was accentuated when I received the bill for the air and train tickets. I do not know how often scientists bring young children and caregivers on work trips, but I am sure that those who do are always doing cost-benefit analyses.

• • •

My negotiations and guilt began even before I left for maternity leave. I was given the option to declare or not to declare my pregnancy. As far as I could tell, declaring meant two things—that my office would be tested for radioactivity (I did not work in a laboratory) and that I would receive maternity leave paperwork a month or two before my pregnancy would become obvious. Until reading our policy, it hadn't occurred to me that one might want to conceal pregnancy status. I chose to tell my supervisor and hide the fact from the bureaucracy because I didn't want to be responsible for my division's having to pay time charges for environmental health and safety personnel to conduct the test, and I wanted to continue to lead the sort of anonymous existence that is not possible when two uniformed staff with a survey meter come to your office. Though one could ask for the test to be performed under cover of darkness, the possible overtime costs did not seem justifiable.

Additionally, I was confronted with archaic maternity leave provisions requiring that I fill out job termination paperwork and surrender my government badge, key, and e-mail account for the duration of my short-term disability leave and unpaid Family and Medical Leave Act bonding leave. A colleague had recently taken leave under these circumstances, and I was surprised that the laboratory had no alternatives to offer her. It was stressful to think of journal manuscripts and page proofs that might get returned to sender, professional society committee work that might not get done, colleagues whose offers of fruitful collaborations might be unanswered, and—perhaps wishful thinking—agency sponsors whose offers of grant money might be unaccepted.

I have never been one of those people who bring a laptop on vacation. But a weekly or biweekly check of e-mail or a trip to the office seemed a reasonable expectation for a scientist on what I thought was a common type of temporary leave. In the end, two creative human resources experts helped me cobble together leave policy loopholes that permitted me my e-mail and badge. With one of these individuals now in a new position and another unexpectedly retired, it may be up to me to maintain the institutional memory regarding how to append leave without pay and leave of absence to short-term disability. (At the time of this writing, our leave of absence policy permits scientists to retain badges and e-mail access.)

· · ·

At this point in the story, I need to provide some background on why I decided to step back from my full-time commitment to science. First, I was thirty-seven when my son was born. Although my career had not yet peaked, I was satisfied that I had made a mark in my field of ecological risk assessment. At about thirty-two weeks into my son's gestation I had won a division-level scientific achievement award. Second, I knew that family and other personal matters were more important to me than work. I had spent months recovering from an automobile accident in which I lost my first child (a daughter who was stillborn). At the same time my sister had spent a year in and out of a children's hospital prior to and following her year-old daughter's liver transplant. Third, I didn't have specific productivity goals; there was no tenure to worry about at my research laboratory. Fourth, the cost of living in my town made it possible to live on a single income.

My decision to work part-time was also informed by advice from many female colleagues. Most were believers in breast-feeding and thought that a half day away at work was as much as mom and infant could handle. Not every mom that I talked to had the opportunity to work part-time (one was responsible for the family health insurance), but most endorsed the idea and had spent at least a few years working part-time themselves. The primary disadvantage regarding part-time status was that if and when I wanted to return to full-time employment, my division director would have to bless the decision, meaning that I would probably have to have grants to cover my salary, benefits, and overhead for an extended period of time; these are difficult to acquire if one doesn't have lots of free time at home or paid time at work to write proposals. One staff benefit that was hard to lose was holiday pay, especially later, when I would be spending my holidays writing grant proposals. On the other hand, I knew two women whose positions in environmental assessment were similar to my own and who had chosen to remain at part-time status well past their sons' high school or even college graduation because of the flexibility. I chose to work part-time, and after four years, have found this schedule satisfying.

My part-time status has made the scheduling of meetings and conference calls, especially with West Coast participants, tricky. Usually I am in the office between 10:00 a.m. and 2:00 p.m. but often earlier or later as well, depending on the caregiver and day of the week. Sometimes I have been embarrassed to tell fellow scientists that I cannot make an 8:00 a.m. meeting. In Sam's first year, I scheduled meetings around the "Mommy and Me" exercise class that we both enjoyed. Though I felt guilty about it, those Wednesdays forced me to become extremely efficient.

Though my husband has been supportive of my decision to work part-time, we have both slipped into the pattern of treating his workday as more important than mine. When Sam is sick or the babysitter cancels or the nursery school has a day off, I am usually the one to stay home—even though my husband is paid for personal days and I am not.

Many scientists down the hall and almost all of my colleagues and project sponsors at a distance have been surprised to discover that I work only part-time. I have managed to write enough papers to fool them, largely because I am a good and efficient writer, I collaborate with productive people, and my type of science doesn't always require much data. A female colleague suggested that I not broadcast my part-time status. I still wonder how my secret might influence would-be project sponsors and collaborators.

Avoiding travel is a challenge for a professional scientist. Early on, I informed my supervisor that I wanted to limit my conference and other unnecessary travel—a request that he supported. In fact, my goal was not to travel at all during the first year after Sam's birth. I did not want to be that woman in the restroom at conferences pumping at every break. Nor could

I imagine leaving my son overnight. But as time wore on, it didn't seem fair to impose the burden of travel exclusively on my collaborators.

I recalled that a friend at a California university had traveled extensively with her daughter even when she was an infant. There was also that colleague who taught me about toddler harnesses. I have now traveled with my son three times for work. The first was a landscape ecology meeting in Las Vegas when he was nine months old. I hired a mature, professional babysitter with a reputable firm (there are lots in Las Vegas because the casinos do not allow children) to watch Sam for an afternoon while I gave my talk and listened to a few related presentations. The arrangement did not go well. The only activities that kept Sam's tears at bay were endless stroller rides around the pool. I missed much of the conference, though a good female friend and mother watched Sam so I could attend a few key talks. I have never again used a "professional babysitter." My second trip was a Department of Defense conservation meeting in Savannah, Georgia, when Sam was about fourteen months old. My husband and I made a vacation of it, and he watched Sam almost full-time for the three days of the meeting. Though I brought Sam to the poster session reception, it was more formal than the poster sessions I was used to and I did not feel comfortable with a child in tow. My third, and least satisfying, professional trip with Sam is described at the opening of the chapter.

My most recent challenges involve obtaining enough funding to pay my salary. Many environmental scientists are suffering under the priorities of the Bush administration. Furthermore, extensive traveling to promote my research is not possible for me. At times I find myself working full-time hours with no salary, writing proposals that have a low chance of funding. This practice is unsustainable, and it affects my mood in all aspects of my life. For the first time, I recently found myself not wholeheartedly recommending part-time employment to young staff members interested in motherhood or second careers.

I now think that a few part-time sources of income (for example, a combination of teaching, researching, and writing) would be ideal to absorb the uncertainty of those lean times in environmental research. But there are many institutional barriers to this kind of income insurance. For example, my own institution's attorneys were slow to approve my getting paid (using my own computer and home) to write articles for a toxicology trade newsletter.

It is unlikely that a different job would improve the balance in my life. Several years ago I interviewed for a few assistant professorships but never got a warm and fuzzy feeling from department chairpersons and deans about the prospect of having children during the race to get tenure. I was asking the maternity leave question well before I had a child because I like to plan for the long term and because I wanted to feel comfortable with potential colleagues and their policies. (A graduate advisor once told me to tell interviewers what they want to hear, but I am not good at that.) Couples within the same university departments have the advantage of being able to take over classes for each other if illnesses or other emergencies come up with the children. Someday this might be an option for my husband and me.

I have enjoyed the challenge of working part-time. However, it is clear that my funding prospects are getting worse, not because of motherhood but because of increased research competitiveness. I constantly question the viability of my job, as do many of my colleagues. I am fortunate to have a position with flexible hours and a flexible work environment. But I have also missed out on a few things. It is possible that if I had not been a part-time staff member, I might have been asked to be a group leader, supervising several other scientific staff members. My part-time status keeps me from getting overhead "burden" funds to pay my salary when grants are not coming in, and working part-time increases the likelihood that I will be asked to share an office (potentially decreasing my productivity). Additionally, my self-imposed travel limitations prevent me from doing the networking I may need to sustain my position or from gaining stature in the scientific societies to which I still belong.

My main career goal has been to conduct policy-relevant science, which I have managed to do, even during motherhood. Knowing many of the challenges of combining motherhood and science, I feel a responsibility toward those women who have left science and want to return. No, I would not preferentially hire a mother over a more qualified candidate. But I will always keep in mind the scientific expertise of full-time mothers I know, in case I have funds for short-term consultants with their expertise, because easing back into the workforce is easier without a gap in one's résumé.

The Eternal Quest for Balance

A Career in Five Acts, No Intermission

Theresa M. Wizemann

Director, Worldwide Regulatory Affairs, Merck & Co., Inc.

PhD, Microbiology and Molecular Genetics, Rutgers University
and the University of Medicine and Dentistry of New Jersey, 1994

It takes the earth approximately 365.2425 days to make one full orbit around the sun. Since it's too difficult to manage a calendar with a quarter day, every four years we throw in one extra. As graduate students each trying to accomplish sufficient research to fill a dissertation, Leap Day seemed to be the only free time my fiancé and I were going to find to get married. And so we did, in the campus chapel (after making sure my cell cultures were fed for the weekend). Thus begins the story of our family and our eternal quest for balance.

As a kid I had strep throat and double ear infections several times a winter. By age twelve, I was reading my mom's Merck Manual to self-diagnose. I finished college with a degree in medical technology, but after working in the hospital labs for a while, I knew I needed something more. Years later my husband and I would defend our doctoral dissertations, one week apart, in microbiology and molecular genetics. We were newlyweds with newly minted degrees and the new problem of having to consider another person in our plans as we forged ahead with our careers. Our fate as post-doctoral fellows was sealed with a kiss, literally, when I gave my husband a quick peck on the cheek as he went into the job fair at our annual professional society meeting. We didn't know that the professor my husband was about to meet had observed us. He readily offered to make some contacts for me as well, sparing us the awkwardness of the "trailing spouse" conversation. We headed to New York City for fellowships at the Rockefeller University and, coming full circle, I began work on *Streptococcus pneumoniae*, a major cause of pediatric ear infections.

Pregnancy during my fellowship was not a problem; in fact, in my laboratory it was almost a requirement. My advisor had been pregnant when I interviewed, and in my first year as a fellow I was the third person to become pregnant. The balancing act began even before the baby was born, with my husband trying to finish a grant application while helping me through contractions. The campus day care did not take infants younger than three months old, so after my six-week leave, I headed into the lab at 6:00 a.m. and left at 2:00 p.m., while my husband worked from 2:00 p.m. until 10:00 p.m. As a result of this parenting in shifts, my husband developed a wonderful bond with his new son, and sometimes I think he is a better mom than I am! Our very fertile lab was a kind of thrift shop for baby supplies. Maternity clothes, baby clothes, and our advisor's baby carriage with removable bassinet all made the rounds. As it turned out, a biomedical research lab focused on pediatric infectious diseases was a great place to start a family. Advice was abundant, and our son's first babysitter was a Swiss pediatrician. But diapers and formula were expensive, and after two years as fellows, we started considering our next move.

It was 1996, and everyone was sequencing genomes. Molecular technology and computer software had taken a giant leap forward, and the race to sequence the human genome was on. In our lab, postdocs were using bacterial genome information to identify proteins or other molecules on the pathogen's surface that might be potential targets for new vaccines. My advisor collaborated with a biotechnology company in Maryland, and after several job interviews for positions in New York that were not quite right, I talked to the company about working there. Thankfully, the hiring personnel were interested in both my husband and me.

I remember learning in freshman sociology that bargaining power is inversely proportional to how much you want something. If I didn't understand it then, I understood it now as I entered my first "real" job. Finding two jobs in the same state, not to mention in the same company, was more than my husband and I could have hoped for, and we jumped at the first offer. There was not much flexibility as far as work arrangements, and panic set in the first time we got the call from day care to pick up a feverish child. Which one of us could abandon that day's experiments? This was no longer our own research, where lost time was just a personal setback (possibly leading to postponing a dissertation defense or not publishing when planned). This was business, and time is money. But the panic was unfounded, and the environment, while not overly family-friendly, was not hostile.

In a small company you do a little of everything, and I developed an interest in the business end of science. In the process, I discovered a field I had never heard of before: science policy. I was fortunate to be able to pursue this interest through an AAAS congressional science fellowship awarded by my professional society. Fellows' stipends are paid by the sponsoring societies. The congressional offices get free scientific expertise for a year, and in return, fellows participate directly in the shaping of federal science policy. As I interviewed at different congressional offices, I was warned that the staff in my office of choice worked long and hard. Hard was not a problem, but long was, and in my search for the best fit I nearly turned down an interview with that office on the basis of this reputation. So when they called to offer me the spot, I didn't know what to do. My ever-supportive husband agreed to be Mr. Mom for the year so that I could make the most of this experience. Unfortunately, with an hour commute to Capitol Hill from our home in Maryland, I would have to leave before my toddler woke up in the morning. In order to see my son at all during the week, awake at least, I would need to catch the 6:05 p.m. train out of the city. To my surprise, the staff director agreed, as long as I could arrange to stay late for the occasional emergency. My year on the Hill, and in that particular office, forever altered my career path, and it is an opportunity that I nearly missed because of fears and misperceptions based on hearsay.

Following the fellowship, I headed to the Institute of Medicine of the National Academies (IOM) to lead a science policy study. With experience and credibility comes confidence, and I felt that I was now in a better position to bargain. Moreover, I had done my own research this time through a network of colleagues and learned about the different types of work arrangements that might be possible. Hours were flexible, and I found I could get a lot done in the early morning before most others were in the office, so I arrived around 7:00 a.m. An unexpected benefit was the personal time I reclaimed by avoiding rush-hour traffic on both ends of the commute. I went in determined to work at home one day a week, and I could telecommute additional days as needed—for example, if I wanted to attend a school event in the middle of the day. I'm happy to say that I have worked at home nearly every Friday since 1999. Though working from home is a wonderful benefit, the downside is that while others can take a day off to care for a sick child or because they are snowed in, I can always get some work in. Lucky me!

Our second son arrived while I was at IOM. Since I was winding down a study and not simultaneously starting up a new one, I cut back to 80 percent time for the last months of my pregnancy. Friends warned that I would just be doing the same amount of work for less money, but I guarded my time carefully. After maternity leave I worked 60 percent time as senior staff on a large, ongoing study. The study director needed help but wasn't sure I would be comfortable working for her since we were colleagues. I saw it more as an in-house consultant job. So, paying only a part-time salary, she got an inexpensive but seasoned researcher and writer, and I got to do the parts of the job I enjoyed—the research and writing—while avoiding staff management, budgets, dealing with study sponsors, and meeting planning.

After working part-time on two short studies, I was ready to direct again and made the jump to a public policy position in the DC office of a major pharmaceutical company. I couldn't persuade them to let me try the job at 80 percent time (that bargaining power thing again—I wasn't willing to call it "nonnegotiable" and risk losing the offer), but I retained the regular Friday telecommuting. I also worked on the train to and from the office, which allowed me to be physically in the office an hour less on the days I did go in. When my husband joined the same company a few years later, we relocated from Maryland to Pennsylvania, where his new lab was, and I did the same job from a different office. Recently, I've moved to a different position within the company but have retained the flexibility I value so much.

What's next? Soon we will have a preteen in middle school, turned loose into a world of temptations at 2:30 p.m. every day. I believe it's at least as important to be available to children during these critical tween years as it is when they are babies. Part-time is probably not a realistic option in my current position, but extreme flexibility could be (such as working afternoons at home on days when he doesn't stay after school for activities). But I'm thinking of taking the big plunge—going it alone as a consultant and writer. It's a calculated risk, aided by the fact that we have steady income and benefits through my husband's job. So for now, I'm putting some money away for a cushion, keeping in touch with all my contacts, and reading everything I can find on consulting and running a business from home. Applying those freshman sociology principles again (now enhanced by management training in business negotiation skills)—a source of power is options, and I'm making sure I have options.

Bringing up a child and bringing up a career are remarkably similar. They are sometimes hard to conceive, and sometimes they catch you by surprise. They are often exhausting, sometimes heartbreaking, rarely predictable, but tremendously rewarding. Motherhood is the grandest experiment. Don't miss out. And remember that you are not alone in your eternal quest for balance.

Reflections on Motherhood and Science

Teresa Capone Cook

Upper School Science Teacher, American Heritage Academy

PhD, Biology, University of North Carolina–Chapel Hill, 1994

The e-mail read, "I hope you will consider writing an essay about your experience as both a scientist and a mother." My first thought was, "OK, no question about my being a mother, but am I really a scientist?" It's true I have a doctorate in biology, but some days it seems as if all that happened a long time ago. I am not currently conducting research, writing grants, or publishing manuscripts. I have even let my membership in AAAS expire. On the other hand, I read *Science News*, occasionally delve into primary sources, and can still converse intelligibly with scientist friends.

When asked to list "occupation" on surveys or medical paperwork, I write "science teacher" (and sometimes "faculty," which sounds loftier, though we all know it means the same thing!). Though I've worked as a high school biology teacher for the past five years, I was certainly trained to become a scientist. My formal graduate work was all about conducting ecological research, nulling hypotheses, thinking critically, writing scientifically, and carefully studying the work of those who had come before me. Along the way I taught labs, recitations, seminars, and in summer science programs, and by doing, I learned to teach. Over those ten years of graduate school my goals changed and then solidified as I saw my life more and more through the lens of my family.

When I began graduate school, I was a newlywed. My husband, Scott, was an electrical engineer with Texas Instruments at the time and willingly left his job so we could both relocate to North Carolina, where I would begin my doctoral work. My master's advisor understood the importance of Scott's readiness to move in support of my career well enough to include a statement about it in the doctoral program reference letter he wrote for me. My own mother had been the first in her family to graduate from college, and she went to graduate school in pursuit of a doctorate in clinical psychology. While she never tells it this way, my take is that she had me and then chose not to complete her program. She had put in the research time but found it difficult to complete the writing of her thesis and ultimately her dissertation. I was determined that I would finish my degree, write my dissertation, and graduate before any children of my own entered the picture.

I defended my dissertation in May of 1994, three months before the birth of our first child. In June I was busily putting the finishing touches on several manuscripts and sending them off. The most important manuscript was a paper on mate choice behavior in stink bugs submitted to *American Naturalist*. The timeline went like this: August 22, 1994, Cara Marie Cook was born; August 24 we came home from the hospital; August 25 I received the *Am Nat* manuscript back in the mail, accepted with revisions. I never completed those revisions, and the most important research from my dissertation remains unpublished to this day. Honestly, I was more concerned about how to breast-feed without six pillows and several more arms than the mate choice of stink bugs. Perhaps this is what my doctoral advisor knew early on when he exclaimed, "I'm wasting my time with you," or something to that effect, upon learning that my husband and I had been so bold as to actually buy a house, rooting us in North Carolina, while I was still in graduate school.

I'm not sure I ever really knew what I would do after becoming Dr. Capone Cook. My mother began her professional career as a clinical psychologist, even without a graduate degree, when my brothers and I were old enough to be at school during the day. That was the first of three different and successful professional careers that she has had. My father was a college geography professor, and the pros and cons of academic life were something I thought I understood. I loved afternoons spent in the stacks of the research library, just about any day in the field, and the pulse of life on a college campus. Of course I also recall dinner table conversations rife with phrases such as "tenure fight," "departmental bickering," "grant-writing deadlines," and "publish or perish," so I knew academia wasn't all wine and roses.

But once I was in graduate school, no one on the inside talked much about alternative career pathways, or even family life, for that matter. The successful path was obvious and predetermined: no stopping along the way, defend your dissertation, publish, secure a grant from the National Science Foundation, apply for a postdoc, publish, complete tenure-track university research, and do some teaching (if you were so inclined), publish some more, and so on. It seemed it was bad enough being married as a grad student, but to have children was unthinkable, at least until you were well established and had a cadre of single, childless grad students working for you in your own well-funded lab. As a graduate student I was highly successful. I loved research and seminars and science, but all the while I was planning my escape. There was something in me that knew I wouldn't be happy or effective competing in the university research world at the same time that I was learning to be a mother, and I did not see a way to step off the train and still be allowed to get back on farther down the road. It seemed possible only for a brilliant, creative few, those who always seemed to think and operate outside the box, and that wasn't me; I am a rule-following collaborator who always prefers to sit in the middle of dinner table conversation. Had those exclusively male established professors been aware of this, would they have accepted me into their rigorous and well-respected program? I do not know. Throughout graduate school I definitely felt that I had to keep my alternative career goals under wraps. And though my grad school cohorts supported and encouraged me, most of them would follow a much more traditional path. Perhaps, unlike my advisors, they have used my career path as an example of a successful alternative when talking with their own graduate students. I hope so.

Toward the end of my doctoral work, I was close to thirty, had been married almost ten years, and was excited about starting a family. Surely the biologist in me could understand this. But what about my ten-year investment? What about my love of biology? When I considered what it was that I liked doing the most, I kept coming back to teaching. I enjoyed it, was good at it, and felt that it was vitally important to train our children and citizens to be scientifically literate. I started to see that my training in how to "do" science, to think scientifically, to be able to communicate clearly through scientific writing, and to understand scientific research was the best preparation I could have had to be able to teach and hopefully inspire my students. As Cara's birth approached, I stopped looking for a postdoctoral fellowship and began to focus on part-time teaching opportunities that would allow me to spend the great majority of my time with my husband and our baby. My alma mater had lecturer positions available in ecology and general biology, and when Cara was four months old, I left her with a wonderful college-aged babysitter for several hours each Tuesday and Thursday afternoon so that I could teach ecology to college biology majors, and then several evenings a week to teach biology to continuing education students. On the weekends, when I would prep my lessons for the following week, Scott got to spend lots of quality time with our baby.

Looking back, I see that each new and different teaching opportunity prepared me for the next one. I felt that God, who had gifted me with this teaching ability and a love of science, was now showing me a way and opening doors for me that I never imagined during my graduate work. I can now say that I have more than twenty years of science teaching experience in a hugely varied set of circumstances—graduate school, colleges, workshops, museums, elementary schools, high schools. Whenever our family circumstances have changed, such as when our second child, Nicole, came along or an out-of-state move was necessary, there have always been interesting, challenging, and family-friendly science and/or education career options available for me. Often it has been the combination of my content knowledge and degrees with my teaching experience that has given me these choices and made me a sought-after employee. The icing on the cake for me is that currently I teach high school biology, zoology, and botany at the same private school that my daughters attend. We get up in the morning, plan for our day, and head off to work and school together. So far neither one of them minds being a teacher's kid, though I have not yet had them in my class as students. I know their teachers and friends, and I see them interacting with their peers and teachers at school in a way that many parents cannot. And I get the summers off!

Am I a scientist? Yes and no. I have been, and I apply the skills and knowledge gained during that time to my life and work every day. I hope that I am training my students to be informed, scientifically literate citizens of our world. It certainly seems possible that I may yet again work as a scientist in the purest sense of the word. Knowing what I know now, would I have chosen to do things differently? Probably not—but I would have been bolder in expressing my goals and career path, I would have proudly stepped outside the box, and I definitely would have finished those *American Naturalist* revisions!

The Benefits of Four-Dumbbell Support

Catherine O'Riordan

Consortium for Ocean Leadership

PhD, Department of Civil Engineering, Stanford University, 1994

Title Nine sport clothing categorizes its running bras by the number of dumbbells needed for support. The dumbbells indicate strength of support, and a four-dumbbell bra is the most you can get. These bras are sold to women with serious curves, who seldom appear in the catalog's action photos or modeling their clothes. I am one of those women whose running routine would not be possible without this level of support. I need to run and bike and swim. In addition to providing health benefits, intense physical exercise releases stress and helps me to cope. The four-dumbbell support enables me to keep going. Likewise, maintaining the balance between my science career and other aspects of my life would not be possible without a good deal of support. Support from family, friends, babysitters, and understanding employers has been essential to working out this balance.

I am a second-generation scientist-mother. When I was young, many of my friends' mothers did not work outside the home, but even forty years ago, there were a few women out there like my mother, a chemistry wiz who planned heart transplants for dying babies as a pediatric cardiologist, took only a three-week maternity leave for each of her own children, and kept six kids at home well fed, injury-free, and successful in school. That she was married to another medical doctor who was focused mainly on his career only added to her sense that she had to do it all herself. She did not have very much support but made up for it by working very, very hard. Of course, a woman had to be twice as smart as a man to prove her worth, she would tell me. Fortunately, for her this was not a problem.

My mother never expected any less of me. My two younger sisters did not enter science or engineering but chose instead to stay at home full-time for more than ten years with their children. They did not have the same confidence in science that my mother and I had. Perhaps they felt that by being home they would give their kids more of what we thought we had missed as children all those hours that my mother was at the hospital saving other children's lives.

But I knew I was going to be an engineer or a scientist or a medical doctor. My two older brothers became engineers, and I could do anything that they could do. So I threw myself into my studies, earning a degree in mechanical engineering. I enjoyed engineering and had no interest at that time in a research career. Four years later, I was working in one of the best jobs ever (monitoring coastal water quality from boats along the coast of Massachusetts) and became interested in hydrodynamic modeling of natural water bodies. After a few years of part-time graduate classes, I quit my job to attend graduate school full-time. Earning a master's and PhD in environmental fluid mechanics, a subdiscipline of civil engineering, built my confidence and inspired me to stay in research. During graduate school, I met and fell in love with Arnaud, a Frenchman, and found a French government fellowship for Americans that allowed me to conduct postdoctoral research in France's national civil engineering school, just outside Paris. The fellowship gave me the freedom to continue water resources fieldwork using instruments that I developed. I worked on boats and in the lab to build a better understanding of natural water systems.

The idea of having children seldom occurred to me during those years, and when it did, it was in the abstract. If I felt secure enough in my professional life, and if my body could do it physically, I might start to think about it. If getting pregnant turned out to be difficult, I did not want the desire of having a baby to become a focus of my life. I knew very few women my age who were both having kids and continuing their science career full-time. Then I married the Frenchman and could see him fathering my children. I was living in France, where the state helps to support working moms with state-subsidized, four-month maternity leave and subsidized child care.

Two babies came two years apart, without my really taking stock of all that was changing; the toughest part of it all was navigating the French health care system and going through delivery in my second language ("Ça me fait mal!"). Even with the extensive support in France for maternity and child care, I found myself unable to remain competitive in research. I could not find enough hours in the day to keep doing the fieldwork, analysis, and publishing, in addition to the search for a more permanent academic position. Frustrated but resigned to leaving research, I began to look outside academia at science program management, engineering positions, and science editing. I knew that there were other satisfying career options for research scientists but was not certain how to move on.

Although I was mainly looking for positions in Europe, I was fortunate when I received an offer to work for a large scientific society of geophysicists in Washington, D.C., the American Geophysical Union (AGU). My husband was supportive, agreeing to move away from family and friends to follow me back to the United States. He had been working for a research institute, and once in the United States he had no guarantee of landing as good a job. Fortunately, within two years he found a faculty position that he loves in fire protection engineering at the University of Maryland.

I completed many different projects for the AGU, from converting membership and accounting database systems to marketing student programs to scientists in the third world to working with the board on governance. More than two years ago I moved into science policy and became the public affairs manager, where I communicated earth science to policymakers and the public. Working in science policy full-time was exciting. The research produced by earth scientists underlies some of the most important issues facing society today: natural hazards, environmental quality, climate change, and natural resource management. Recently, I moved on to another science nonprofit, the Consortium for Ocean Leadership, to manage U.S. participation in scientific ocean drilling where I also seek new opportunities to reach out to teachers, policymakers, and the public. Despite having left research for an "alternative" career, my work advances science and is fulfilling.

I like to think that the hardest years of balancing, establishing a career, and dealing with ear infections and nighttime bottles are over. But I find that my kids need me even more now at eight and ten as they ask questions and begin to navigate the world on their own. I have discovered that support helps: from my husband, babysitters, and friends. My positions in science management and understanding employers have given me enough flexibility that most of the time I can rush home for my children when they really need me. We have been able to maintain their native tongue by enrolling them in a French school. To make this work, we seek support from bilingual babysitters who help with the kids' homework assignments. By engaging help from others, I have been able to keep my career on track, while the four-dumbbell bra helps me to stay fit enough to compete in a few triathlons every year. With some hard work and adequate support, I try to keep it all in balance.

Extraordinary Commitments of Time and Energy

Deborah Harris

Physicist, Fermi National Accelerator Laboratory

PhD, Physics, University of Chicago, 1994

In January 2005 Lawrence Summers, president of Harvard University at the time, gave a provocative speech about why there are so few women in "top positions" in science and technology. His remarks were off the record, but he offered the hypotheses that there are "innate differences" between men's and women's brains, that top positions require "extraordinary commitments of time and energy," and that "few married women with children were willing to make such sacrifices." I was dumbfounded. The "innate differences" issue was so completely absurd and has been time and again disproved that I didn't even bother to get annoyed about it. It was the second comment that blew my mind. Who knows *more* about sacrifices and extraordinary commitments of time and energy than parents of young children? But we figure out a way to make these sacrifices and not lose the core of who we are. If we're successful, our demanding young children slowly turn into actual people with their own lives and interests. My ten-year-old is already much less demanding than my six-year-old, and unless I hang out with my thirteen-month-old nephews, I forget how much easier my six-year-old has become. If we're even more successful, we survive the ordeal having learned something. What most people don't realize is that what we learn as parents can be applied to our jobs as physicists and vice versa.

The first parenting skill I learned as a mother of two and a full-time physicist was how to work under conditions of shifting sleep schedules. I had my first child in the middle of a data-run while I was a postdoc. When you work in a field that depends on particle accelerators being up and running, this is a very hectic time that requires around-the-clock surveillance. While I was pregnant, friends and coworkers expressed concern about my working the midnight to 8:00 a.m. owl shift. The absurdity of those suggestions hit me only after I gave birth. In fact, owl shifts are much easier than taking care of a newborn child. When the shift is over, you can sleep for eight hours continuously if you like, and owl shifts last for at most a week at a time—not three or four months!

The way one is treated as a parent in physics (as in most professions) varies as much as the people one works with. For women this variation shows up as early as when one negotiates maternity leave. When my son was born, I was a postdoc for the University of Rochester, had a great boss, and was able to take off three months full-time and three months half-time. Four years later, when my daughter was born, I was working for Fermilab and hadn't racked up enough vacation days to have three months' paid leave, regardless of how supportive my boss at the time was. I know of postdocs at other institutions who got only six weeks' unpaid leave in spite of the Family and Medical Leave Act! But six weeks of unpaid leave sounds like a party compared with a friend's experience—she has left the field and is suing her ex-boss after she was *denied* maternity leave and had her salary severely cut after she had a baby.

I have seen this variation in helpfulness toward families persist at least as far as my child's grade school years, and I would imagine it's there for the duration. Some laboratories have on-site day care, while others have none. Some laboratories have on-site summer camps and "gap care" programs for school-aged children during school holidays; others have none. Some experimental collaborations (which in particle physics often consist of over a hundred physicists) agree to have meetings only over the weekends to accommodate faculty teaching schedules, while some agree to have meetings only during the week. Some bosses arrange meetings or seminars that start at 5:00 p.m., when many parents leave to pick up children, but some do not. Fermilab has an excellent on-site child care facility that I have used for the past ten years. Part of my feeling that I can be an involved parent and a physicist comes from seeing others in the same predicament every day, and part of it comes from the fact that some of the women who have watched my children are also the women who watched my bosses' children.

Some aspects of the life of a physicist are actually conducive to being an involved parent. First, because we don't punch a time clock (and because we sometimes work such strange hours), we can do things like volunteer in our children's classes or come in for special plays or concerts that happen during the school day. I have been volunteering roughly one morning a month for the past four school years, and while I get to know my son's classmates and get a feeling for his classroom environment, I am sure my overall research productivity has not waned because of this. It's ironic, since the perception of a job in science is exactly the opposite. There's a block party in my neighborhood every year, and every year I am asked by at least one new person, "So are you working full-time?" with a note of a surprise in his or her voice. I used to think the surprise was in reaction to my having chosen to work full-time with two young children, but now I realize it may simply come from not knowing how I could cope with the demands put on my workday by school-aged kids. All I can say is that it's a lot easier in physics than in many other full-time jobs.

There are still many opportunities for mothers to feel guilty regardless of how flexible their employment is. Halloween, for example, presents a yearly opportunity for guilt. Some parents (even physicists!) go all out and make gorgeous creative costumes, and some (like me) head over to Costume City. Fortunately, my kids don't seem to mind the Costume City routine, and I just bite my lip to keep from commenting about the inane stereotypical costume options! But as skimpy as I might be about the costumes, I always spend a few hours each year carving pumpkins with my kids.

Raising children forces you to make choices and prioritize. Do I spend a few hours making a costume after my kids go to sleep, or do I log on to work instead and then spend a few hours the next day carving pumpkins? Do I work through a quick "back of the envelope" calculation whose results will pave the way for a completely new experiment, or do I spend my time trying to use a detailed software package to do the same calculation, much more slowly but more precisely? In both parenting and research, there are always many more tasks to do than time to do them, and you just have to figure out which tasks will be the most effective.

Raising children forces you to pick your battles. I would not be caught dead in the color combinations my kids choose to wear to school, but on the other hand I will not let them leave the house without making sure they have had breakfast and have brushed their teeth. As much as I would like to convince the three experimental collaborations I am on never to schedule meetings over the weekends, I know this will not happen, so instead I struggle to get the experiment for which I am project manager funded.

Raising children in a two-career family (which applies to all physicist-mothers I know) forces you to realize that more than one person can do most jobs. My husband does a little more housework than I do, but I do much more with the kids than he does. However, I travel for work, probably spending about four to six weeks away from home each year. It is during these times that my husband really gets to know his children, what they are doing and concerned about, and usually when I return he is full of information to share with me about the kids (most of which I already knew but don't confess). As project manager for a new experiment, I have seen technicians, engineers, students, and faculty all contribute outside of their original job descriptions to get the experiment built. Just as any one person can do more than one job, any task can be achieved by more than one person.

Unfortunately, the negative assumptions about how scientists fare as parents (and how parents fare as scientists) get far more publicity than the reality. One assumption you often read in the press is that every minute of research counts, that if you lose three or six months in a row or even three to four hours in a twenty-four-hour day because you are caring for your children, then your physics (and funding for your physics) grinds to a screeching halt. However, the same press loves to point out that every great physicist had outside passions, from Newton to Einstein to Feynman. One day perhaps the same press will make the connection that perhaps what made these scientists great was *not* that they were doing research every waking moment of their lives but that they applied the lessons from their lives to their research.

Finding My Way Back to the Bench

An Unexpectedly Satisfying Destination

A. Pia Abola

Research Fellow, The Molecular Sciences Institute

PhD, Molecular and Cell Biology, University of California, Berkeley, 1995

I was never one of those little girls who dreamed of the ideal wedding, the ideal husband, and the single-family detached house in the suburbs complete with a white picket fence. I am the daughter of educational immigrants—two Filipinos who came to the United States for graduate school and ended up staying here where the opportunities are. My upbringing stressed hard work, striving for excellence, and the importance of education. My dreams were of being a successful scientist, although it was not until college that I settled on a specific discipline: biochemistry. Since I am Filipina, my upbringing also emphasized the importance of family and the joy of children. When my cousins came to visit from the Philippines, I was always jealous that there were four or five children in their families whereas we were only three. Growing up, I always knew I'd go to college and then graduate school, get a job, get married, have kids, and keep working. It never occurred to me how difficult it would be to have it all, that I would initially choose family over career, or that I would not only find my way back to research but enjoy it so much more than I did before I stayed home with the kids. There may be some who believe that your brain decays when you are not doing science, but I know that the time I spent at home caring for my children has made me a much better scientist. I am more efficient with my time and better at planning and prioritizing; I am more pragmatic and goal-oriented; I am humbler and better at dealing with and overcoming my own shortcomings and those of others; I am better at negotiation and compromise; and I am much better able to tolerate the tedium and myriad little failures that accompany work at the bench.

It was during graduate school that I started to comprehend the realities of how hard it can be to combine a scientific career with motherhood. I watched my older friends—mostly postdocs—struggle to balance the science and family equation. Those of us without kids would gather every now and then and have fretful conversations about how to solve this equation in our own near or distant futures. To the best of my recollection, these were conversations I had with other budding female scientists; the male voice was nonexistent, either through exclusion from the group or lack of interest in the discussion. The main work-life issues we worried about were (1) the two-body problem, or how to find a job in close geographic proximity to your spouse; (2) the best time to start a family—during graduate school? postdoc? pretenure? posttenure? (3) how to manage issue 2 if you couldn't satisfy issue 1; (4) how to find enough hours in the day to maintain a brisk research pace while dealing with morning sickness, babies, pumping breast milk, and the vagaries and expenses of child care; and (5) how to do all of the above while maintaining the respect of your colleagues. The only thing that was clear from these discussions was how murky the future appeared. It all seemed like various combinations of compromise and luck. It was especially disheartening to look around at the successful female faculty who, for the most part, were either childless or divorced. A recent discussion I had with an old colleague and friend highlighted the advantages of joint custody of her child for her scientific career—every other week she had the freedom to stay in lab as late as she needed to.

I was lucky enough to go to graduate school in the San Francisco Bay Area, a place rich in opportunities for a biochemist. When Ronald, my husband, was still a graduate student in linguistics, I was ready to move on to a postdoctoral position. Fortunately, it was fairly easy for me to find a position within somewhat reasonable proximity to my husband's school, though it still required a two-hour commute each day. After all the worrying about the best time to start a family, bearing in mind both academic and biological clocks, I decided that the optimal time would be during my postdoc. One and a half years into my postdoc I gave birth to my first child, a son. I worked until my due date, continuing my two-hour daily commute.

My son came a week after his due date in a traumatic birth experience that included a failed induction, C-section, and subsequent anaphylaxis, resulting in a twenty-four-hour stay in the ICU for me. It was one of those experiences where the struggle makes the victory all the more precious. I stayed home with my baby for his first twelve weeks and then returned to work. My amazingly supportive husband and I were able to work out a schedule where we had nearly distinct work schedules so that our son needed child care for only eighteen hours each week. In retrospect, I think this approach was great for our child but hard on our marriage, as we rarely saw each other. Though we moved closer to my work so that my commute time was cut roughly in half, I resented every minute I spent stuck in my car and away from my baby. For the first year I cried every day on my way into work, and my heart was simply not in the research. I hated being so divided.

It was at this time in my career path that I was supposed to be putting together a research program so that I could set up my own lab as a principal investigator. My heart wasn't in it. I couldn't imagine either asking Ronald to leave graduate school and follow me to another part of the country or living apart from him and raising our child on my own until Ronald completed his PhD program. I also wanted at least one more child but didn't want to have to go through the pain of separation again. I agonized about my future. Through my inaction, I gave up on the idea of an academic position and never put together a research plan or initiated the long process of applying for academic positions. I found a project management position with one of the genome sequencing and microarraying centers and thought that this move away from bench work would allow for a more flexible schedule, easing the logistical difficulties associated with having another child. I thought that perhaps after my second child was born I would be able to transition into a project management position in the biotech industry. I had what seemed like a workable plan. And then I became pregnant with twins.

I found out about the twins pretty early in my pregnancy, after an ultrasound for minor complications at eleven weeks. I was depressed for the first three days. How were we going to afford child care for three children, especially when I, the primary income earner, was getting paid at National Institute of Health (NIH) pay scales? How was I going to manage all the logistics of child care and expressing breast milk for three children? How much worse was the pain of separation going to be? I mentioned earlier that the future seemed dependent on compromise and luck. I'd already compromised; now was the time for luck. Ronald's department had been trying for the better part of a year to hire a linguist who was also proficient at computer programming and database management—a tough hire given the then out-of-control dot-com boom. They turned to Ronald who had programming experience, and asked him to apply for the position. He had to put his PhD on hold (another compromise), but his new salary would be as much as or more than he and I had been making combined, and I suddenly had the option to stay home. Never before had I contemplated making this choice. It was completely shocking, it was frighteningly liberating, and for a long time it made me feel like a total and utter failure.

I felt that I was letting everyone down, betraying my graduate advisor, my postdoc advisor, and my undergraduate advisor. I felt I was letting down my fellow women scientists and my fellow scientists of color. But I really couldn't see any other way to make this work and stay solvent and sane. I ended up staying home for five years. I was ready to go back to work after two, but it took me a while to decide what it really was I wanted to do and figure out what I needed to do to get that position. I talked to many people and decided to go back to my prehiatus plan of being a project manager in industry. It really is true that getting a job when you don't already have one is a lot harder than getting a job when you do. Add to that the questions about my dedication to work, given that I'd already placed my family first, and I couldn't even get my foot in the door of any biotech companies. Fortunately, my network of friends and old colleagues that remained in the Bay Area included a woman who worked for a private, nonprofit research institute. She said that money was tight but that if I could secure my own funding, there might be a place for me at the institute where she worked. She pointed me to the NIH supplemental grant program to promote reentry into a biomedical research career.[1] I interviewed at the Molecular Sciences Institute and obtained a position dependent on funding, which I subsequently applied for and was awarded. As I write this, I have been back at the bench for one year and two weeks and am incredibly happy.

I am grateful to NIH for its program to promote reentry into biomedical research careers. This opportunity was an award, but I also view it as a gift. The science is stimulating and fun and satisfying. I have a second chance to try for an independent research career.

1. PA-04-126, http://grants.nih.gov/grants/guide/pa-files/PA-04-126.html.

My children are all in elementary school, and, as my postdoctoral advisor told me when I first revealed my decision to stay home, "They need you less as they get older." Her message was basically one of hope, and I didn't really understand it until I experienced it firsthand. It gets easier as the kids get older—you do get your life back. I'd like to second her message of hope.

Mothering Primates

Devin Reese

Science Curriculum Specialist, Teacher, and Naturalist

PhD, Integrative Biology, University of California, Berkeley, 1996

The motivation for the choices I have made about my science career and motherhood really came home to me in a recent job interview. An astute interviewer asked me, "Are you sure you are ready to take on this job after having been home with your children so much?" I surprised myself by answering, without pause, "I am sure. I am a very ambitious person. That ambition that drove me to devote myself in such a focused manner to my children during their early years now drives me to ramp up my own career again." Since that day, in thinking about the surprisingly cogent answer that sprang forth from me regarding what I had thought was muddy terrain—my own desires vis-à-vis the balance of work and family—I have had some other insights.

My career as a scientist began very early in life. It was born in the woods of Glover-Archibold Park behind my childhood home in Washington, D.C. There I explored alone, searching for eastern box turtles and watching how they reacted when I set them in the rivulets of water running down from the hillsides. Counterbalancing my otherwise urban life, those woods were a leafy oasis. I loved their embracing beauty and the complexity of all the animal lives within them. In the winter, I lamented that the bare silhouettes of the park's trees revealed the boxy hospital buildings behind them and temporarily reduced my Eden to a patch of trees.

By the time I was a teenager, I had read and reread John C. Lilly's books on dolphin communication and was awestruck that one could make a career of studying animal behavior. I was yearning to study animals and chose to work in a reptile laboratory for my high school senior project. From there my experience with animal behavior deepened, and I went on as a Harvard student to create a major in ethology. Classes on the behavior of everything from social insects to primates left me spellbound. I decided to study ants. Many an hour I spent bent over *Aphaenogaster rudis* colonies, recording their movements to and from the baits set out to assess how food source size and distance from the colony interact to determine recruitment. My undergraduate thesis was mentored by E. O. Wilson and contributed to my graduating with honors.

My pure affection for watching how animals conduct their lives grew tendrils into broader disciplines, such as ecology and conservation biology. I conducted my graduate work at UC Berkeley and served as the principal investigator on a study of the effects on western pond turtles of damming the Trinity River system. As part of an effort by the Bureau of Reclamation to determine flow regimes for the basin, I became involved in the applied science of dam management. Still, my original interest in animal behavior reigned, as I reverently followed radio-collared turtles through the woods of the Trinity River Basin. After walking for long, silent stretches through sometimes snowy terrain, I was thrilled to hear the muted beep-beep-beep of the transmitters.

Broader still, my postdoc with the U.S. Agency for International Development through an American Association for the Advancement of Science Fellows program launched me into the world of international watershed management. I moved to Panama to help build the capacity of the Panamanian government and private sector to protect the integrity of the Panama Canal watershed. This diplomatic position found me leading a collaborative planning effort to develop and adopt watershed health indicators. It was during this period of my broadest environmental work, the most far-reaching yet bureaucratic, the furthest removed from the original walks in the woods that had drawn me to this career, that my first child was born. And her entry into my life brought me sharply back to my roots in the observation of animal behavior.

Suddenly I was the quintessential primate mother: doting, protective, territorial, risk-averse, nurturing, and doggedly focused on her well-being. My hormones willed me rapidly to the task of caring for her, closing other windows of opportunity, whether it was an evening out or a new project at work. I negotiated my work with USAID to be half-time and spent the rest of my time enmeshed in the joyful and terrifying experience of being a mother of the species with the most helpless babies known on this planet. My hormones did such an effective job that I had no desire to be anywhere but wherever I could best assure her comfort and fill her with the milk that I so miraculously produced and that she, equally miraculously, appeared addicted to.

Many questions emerged for me during what proved to be the most intimate encounter of my life with that always-beckoning realm of animal behavior. How did she know to nurse? How was my body able to function on such broken sleep? How could primate mothers in the wild survive such a demanding and exhausting job? Who in the troop helped them? Why did modern humans put babies into their own rooms to sleep when a wild baby primate would never be left that far from the group? What were the impacts of that sort of early isolation during the night? Did it confer upon *Homo sapiens* that extraordinary degree of independence later in life? How much was my baby's early food selection determined by food recognition from my own diet during pregnancy?

As she grew and my two other children were born, I became interested in longer-term questions about the development of various behaviors, such as, To what degree is a child's attachment to a mother determined by whether he or she was nursed? How does an innate aversion to unrecognizable foods drive children's characteristic narrowing of diet during preschool years (which may correspond with the reduction of parental guidance in food selection in the wild)? This wealth of questions has kept me deeply engaged in the care of my children for more than six years. Whether they are eating, sleeping, running, or interacting with others, I invariably watch them with curiosity through my primate lens. When the kids fight (to which I respond as a parent is supposed to—"Please don't hit," "Please speak nicely"), I think with relief that at least they are not likely to kill each other like sibling hyenas.

My science career has undergone an interesting evolution during these intense years. With so much of my energy and focus directed to the children, I have transformed my professional activities to nurture their intellectual development; like a blanket, I have wrapped my work life around them both temporally and geographically. I have worked as a contract science instructor, developing and administering curriculum in English and Spanish, for the Child and Family Network Center down the street. My own children piloted the curriculum, raising mealworms and creating fossil footprints and planting flowering bulbs, before it went out to my target group of four-year-olds. I periodically teach animal ecology and other science courses as a contract instructor for my daughter's elementary school; she becomes my colleague as well as my student, contributing ideas and inspiration to the lesson plans. And I work as a weekend naturalist at Huntley Meadows Park in northern Virginia, where my children come to explore the wetland for sleeping beavers or other special discoveries reported to me by the visitor public. My work is not abstract or untouchable to them; rather, it is an extension of our family's collaboration in the study of natural science.

Lest I paint a more ideal picture than the grainy, contoured truth of my recent career trajectory, let me say that there have been bumps in the road. This transformation of my career has brought a reduction in income and in status as measured in the professional world. Earlier in the transformation, when asked that age-old cocktail party question "What do you do for a living?" I was prone to mumble some lame answer such as "*Just* take care of my baby" or "I work a *little* bit but not in a fancy job like I had before." Of course, people smile sweetly and say, "Oh, but taking care of a child is the *most* important work," to which I nod while trying to dodge the conversation. At times I have found myself bewildered in a playgroup where mothers are discussing where to find the best prices for home products or manicures.

Sometimes I have longed to return to the acclaimed, higher-paid world of a full career but have invariably run into my own refusal to enter a milieu where my dedication to my offspring could be perceived as a liability rather than a manifestation of my strongest characteristics. A couple of years ago, I interviewed for a job with USAID in the very Environment Center where I had previously worked. Hoping that it might have part-time potential, I struggled with when to reveal that half of my time was to remain committed to my young children. The interview seemed to go well, but my admission that I wanted a part-time position dampened an otherwise lively discussion. My former colleague asked whether I was aware that there was a full-time day care facility in the basement, which of course I was.

As my children's school days grow longer, I find myself increasingly available to my career, still in tandem with them. We seem to be inextricably meshed in a tapestry of care and growth and learning together. With more of my professional self to offer, I have just accepted a position with the National Science Resources Center to work on K-12 science curriculum; while part-time, it launches me back into the downtown world of work with a national scope. My children are older, and I am both more grounded in my commitment to them and more available to the rest of the community. It has been a remarkable journey. I hope to continue to look back with appreciation for what I was privileged to offer to them, what they offered to me, and how our collective science work contributed to enlightenment beyond our own small troop.

Finding the Right Balance, Personal and Professional, as a Mother in Science

Gayle Barbin Zydlewski

Coordinator, Cove Brook Watershed Council, Winterport, Maine; Research Faculty, University of Maine, School of Marine Sciences, Orono, Maine

PhD, Oceanography, University of Maine, 1996

I made a conscious decision to become a biologist when I realized in high school that this was a field that truly challenged me. While most other sciences came easy, it seemed the concepts of biology and ecology were a little more abstract, interesting, and harder to grasp. In my early years as an undergraduate I easily made my way through calculus, chemistry, and physics, and though professors in those fields tried to convince me to change my major, I held my ground. While in college, I never once noticed that all my professors were men until one day when a biology faculty member made the statement that he was unsure why they were bothering to educate women because women were just going to get out of school and pop out babies! Well, there was a wake-up call. Though at the time I was not married and did not have expectations of getting married or having children, the remark still stuck with me. Here was another challenge in my life.

As a graduate student in the early nineties I began to notice the low ratio of women to men on science faculties. However, it was obvious that among graduate students the ratio was quite high. At the time this was encouraging. I figured, "hey, times are changing—when we all get our degrees, the upper-level ratios will turn!" Hah! I continued on the typical path to academia, receiving an MS and PhD and completing a postdoc. Upon receiving each degree I was highly driven to get to the next. It seemed an obvious progression, and since I still had not met Mr. Right, I plugged along toward what I always figured would be the ideal job: a position at a university or college.

I met Mr. Right (Joe, also a fish biologist) while conducting my postdoc. Now you might be thinking, "Oh no, the death knell, a two-career family with careers in the same field!" We were both completing postdocs and applying for jobs (academic, federal, state, whatever we could get with money attached) with the understanding that we would go wherever the first job was secured. Well, we both interviewed for the same government position. The position was withdrawn before anyone was hired, but it was later re-opened and we both received calls asking us to re-apply. In the meantime we got married *and* I became pregnant. Again, we both interviewed for the position thinking it would be great if the position was offered to him and I could stay home for a while with the baby. Instead, I was offered the position and visited the laboratory, on the West Coast, four months pregnant! Since this was an offer in hand, we could not turn it down, and Joe agreed to stay at home for one year with our baby.

When I was seven months pregnant, we moved across country. I started a full-time government job and soon had our son, Orion. Since I had not accumulated any sick leave, my supervisor worked with me so that I could be paid during my six-week leave. I went back to work when our baby was six weeks old. I was fortunate enough to be able to work full-time and nurse our child. Joe brought our son in every day at lunchtime. Although at the inception of my position I did not have a laboratory or any employees, over the following year I established a primarily externally funded laboratory with several full-time employees and a graduate student at the local university. However, time spent with my family was at a premium. I was traveling a bit and definitely working more than forty-hour weeks.

After eighteen months at home, Joe felt the need to get back to his career. He had volunteered a great deal of time with another government office and eventually secured a temporary position with pay that was much less than he deserved. During this time he, too, established a program with several employees and good funding. When he was promised a permanent position with a pay increase that was not materializing, we started to realize that we needed to think about other options—not only for his career but also because both of us were working well over forty hours per week. Orion, who was now in day care full-time, was getting less and less of our attention even though Joe and I shifted our work schedules to minimize his time in day care, which came at the expense of time with each other. At this time a colleague on the East Coast (the one we call home) pointed out a position opening. I mentioned to my supervisor the possibility of my husband's applying for that job. She immediately took action to try to get Joe into a permanent position at his current post so we wouldn't have to move. This was a family-friendly gesture, but while my supervisor was responsive, others were not, and a permanent position was not offered in time.

So Joe took the position on the East Coast and I opted out of my career—after four years and a promotion to a high-level position. The job offer was perfect, and the location was close to his parents and my mother (three to four hours' driving distance). The decision was really mine, since it would mean giving up my ideal job for an unknown situation. I decided it was the best thing to do for our family. After all, we were moving to familiar territory: his offer was at the University of Maine, where I had received my PhD. When we arrived, I decided I would stay home full-time with Orion. Before the end of the summer, it was obvious that I needed to do something more than stay at home, and Orion (then four) really needed to interact with other children. By the fall we had enrolled him in preschool, and I had visited the current director of the department where I had received my PhD. Joe's job was just as great as we had expected, and Orion enjoyed his new preschool and friends. I was provided a soft-money position as research faculty with office and laboratory space. Was I opting back in?

I worked my tail off for one year and secured funding for some small projects with approximately one to two months of salary. I completed the projects and continued to write grants for funding, volunteering my time six to seven hours daily at the university. The following fall Orion went into kindergarten, and I continued to volunteer my time at the university. At this point I had secured money for a couple more projects, including salary (for about six months) and money for a graduate student. I tried very hard to connect with other faculty members, but it is difficult to connect when you have to return home to meet your child's bus at the same time that your colleagues are ready to get out of their offices and talk (or when faculty meetings are scheduled in the late afternoon). In some arenas I was merely my husband's wife. I also applied for several jobs in the area: one federal, one state, and one academic. When I did not get the federal or state job and then did not get interviewed for the university position, my confidence plummeted. Another disappointing twist was that Joe was told that he and I could not apply for grants together. After days of arguing our credentials with government human resources representatives, he was told that it was not an option and we could not put both of our names on anything that required our signatures. This made me frustrated by the system and society's failure to recognize that multiple aspects of one's life can (and likely will) intersect—be it husbands and wives working in similar fields or parents having children and still wanting to work (a concept that has mystified many of my male and female colleagues!).

This is when I started realizing that maybe all this hard work really wasn't worth it after all, especially if Joe and I could not even work together (the foundation of our relationship). The attempts to maintain my career were all still coming at the price of time spent with our family. To meet what I felt were "academic" expectations, I was waking up a couple hours before everyone else to squeeze in work and then was always at the computer when at home. I started looking for a female role model in a similar situation. There weren't many. I could identify only a few full-time female faculty members with children, but they seemed unapproachable, and I did not want to take their time.

In my own department, however, one of the female research faculty was a mother. She and I could barely find time to talk, since when we were on campus we needed to pack our time full with research and then get home to our children! But I heard the same things from her—that she wondered whether it was worth it and that she was cutting back her university time to be at home. One encouraging aspect of research faculty positions at the University of Maine is that some individuals, including this colleague, were provided an "agreement." The office of the vice president for research matched 40 percent of salary brought in on grants. Although this possibility was mentioned when I arrived, the time was never right to request such an agreement for me. After two years I raised the issue with my department chair again, and unfortunately, the time still wasn't right, politically. I then visited the dean, described the volunteer work I was conducting at the university, and argued that I at least deserved the same deal that some of my colleagues had, adding that I would prefer a part-time position. He said that if I had a National Science Foundation (NSF) grant on my slate and more teaching experience at the university (which would come at the cost of my research, the performance measure of my position), then I would be in a better position. I immediately decided I needed to get an NSF grant but then thought about it again. If I had a large NSF grant, would I have any time for my family?

I was ready to give up on the university and find other options. But I knew I could not sit idly without doing something outside the home, and besides, we needed the money. While I am grateful for the opportunity provided by the chair of my department, the continual volunteer time for little return and recognition is not very rewarding. Recently, an alternative opportunity fell into my lap. A local watershed council I have been involved with had hired a coordinator who quit the job within a month. I hesitated to offer my services. It is not what I trained to do all these years, but I am now doing this job and finding it very fulfilling! It is part-time, at home, without the pressures of academic expectations that I am unsure I can fulfill right now, especially in a volunteer capacity. I still maintain my laboratory and graduate student at the university while working twenty hours a week for the watershed council. Perhaps naively, I am hoping that some day I can opt back into that academic setting more fully. In the meantime I am being innovative in sustaining commendable work that is satisfying and makes a difference for a natural resource while keeping one foot on another career path I still value.

It is frustrating not to find mother mentors in the academic setting, but it is important to provide a role model for young graduate students. Just recently I was given a boost by a female graduate student about to be married. We were discussing the fact that my husband and I were dragging our son to a professional meeting because we both wanted to go and we didn't want to leave him at home with someone else. She said to me, "It's good to know that it can be done!" I pessimistically said, "It's not easy," but her words gave me the ambition to continue to show a good example, that part-time engagement in the academic setting can work. The key is getting that type of position accepted within the social regime of academia.

There are many negative sentiments about women's abilities to conduct certain jobs and get to upper-level positions. If we focus on this issue only, however, then the options that enable admirable career opportunities, including opting out or partially opting out for a while, will continue to be overlooked and not highly regarded. I will continue to find a way to maintain myself as a person and be a mentor to developing female scientists. Perhaps this means helping the academic community truly accept the concept of part-time and alternative career paths in science.

What? I Don't Need a PhD to Potty-Train My Children?

Nanette J. Pazdernik

Cellular Biologist, Writer, Adjunct Professor, La Leche League Leader

PhD, Molecular, Cellular, Developmental Biology and Genetics,
University of Minnesota, 1996

Oh, to be naïve and young and dedicated to my career above all else. . . .
That is how it all started. I grew up in a small Wisconsin town and went
to a small school and saw the same people day after day after day. Fascinated with the nature that surrounded me, I loved to watch how the world
changed ever so slowly each passing season. Where and how did the fish
survive the winter? How did the trees survive? How did I manage those
long, cold winters?

School! That is how I survived the winter. I loved learning; I loved the
regularity and rhythm of school. I excelled from day one and never had difficulties in any class all the way through high school. My true passion was
for biology, and I graduated at the top of my class with a love of nature, the
environment, and science.

I loved learning so much that going to college was not optional. I chose
the college with the best financial aid package and the best size for my small
town background. Lawrence University in Appleton, Wisconsin, is a small
liberal arts school with fewer than 1,200 students. The professors teach the
classes, there is a little bit of research—though not the kind likely to be
published in *Science* or *Nature*—and there are plenty of new faces for someone from a town of just a thousand people. I was in heaven!

What a change it was to go to college. For the very first time I had to work hard to do well, but I loved the new challenges. I particularly loved laboratory work and decided to do a senior honor's thesis, for which I learned basic molecular biology techniques by reading different protocols, trying different experiments, recording my results, and thinking about technical improvements. I wrote a thesis on my research, selected a committee before which I defended my thesis, and ended up achieving magna cum laude for my work. Not the typical course of study for someone attending an undergraduate college!

Looming on the horizon was life after college. I liked my research project, yet I still wasn't convinced I would enjoy a career doing science. I didn't know if I wanted to apply to graduate school, but I took the GRE just in case. Rather than attending graduate school immediately after college, I thought I had better get some experience conducting research every day. I took a laboratory technician job at the Medical College of Wisconsin, where I learned how to culture heart cells from chicken embryos. I watched with awe as all the little beating heart cells grew together and worked in unison to beat as one. I did enjoy doing experiments every day, so I applied for graduate school the next fall and was accepted by the University of Minnesota program for molecular and cellular biology.

I started my graduate career with such excitement, planning to be a scientist and a teacher. I was going to devote my life to teaching others the mysteries of science and inspire my students with exciting lectures, experiments, and an enthusiasm for learning. I survived the long, tough first few years of constant study and work in the laboratory, and although I found it difficult, I persevered. Interestingly, I found that people attending graduate school were older and married, and some had even had previous careers. As a student with so little life experience, I found everyone else so interesting. During my second year of school, I met the proverbial Mr. Right in genetics class. Dave was getting his PhD in the agronomy program and had also grown up in a small town. We started dating, and I never looked back.

Dave and I married during my last year of graduate school, and suddenly we had to find employment in the same place at the same time. Initially we found compatible employment opportunities. Dave found a job outside Indianapolis, a location close to the Indiana University School of Medicine, where I found work as a postdoc doing medical research on cell signal transduction. We lived halfway between the two jobs. He made about triple my postdoctoral salary. I worked and worked and worked, mostly on futile experiments that produced unpublishable results. Finally, after getting very exciting and very publishable data from a side project, I worked hard to write up the results. After three years of working my tail off for so little salary, I published one paper, which was not enough to compete for a faculty position. That was the year I turned thirty!

This is one of those topics not discussed with women who enter graduate school in science: women in science have to be the most productive during their best childbearing years. I know that one can wait until an older age to have children, but I think it is a lot easier to be younger. My thirtieth birthday was a very sobering time. I had done everything for my career, but did I want to forgo a family? I didn't want to be alone on my deathbed with only my journal articles to keep me company. My husband's aunt had chosen not to have a family. She had a wonderfully glamorous life with exotic trips, multiple husbands, fancy clothing, and a beautiful house. But she died just two years before my thirtieth birthday, and the end of her life was not glamorous. She cried and cried about the extreme loneliness. She begged Dave's mother to "share" her children so she would not feel so alone. Well, the big picture is always much more sobering than the small picture. Finally, I got pregnant! And this is where the story really begins.

Dave completely agreed with my decision to stay home. We didn't need my pittance of a salary. I was so incredibly tired from working fifty or more hours a week on futile experiments, I thought having this baby would free me from my job. (For those readers who do not have any children, take note: a baby requires more than fifty hours a week of physical labor and is definitely much more emotionally draining than any job. See my naïveté?) I planned on returning to work after I had finished breast-feeding my baby, so my hiatus was going to be temporary. I also knew that if I left my job, I could find another postdoc position after a year of taking care of my son, but this is where dual careers can be difficult. Dave's company was bought out and consolidated with two others. His job near Indianapolis evaporated into thin air, and luckily he was offered a job in Mt. Vernon, Illinois. Mt. Vernon is in the middle of the southern half of Illinois: far from Chicago, far from the capital of Springfield, far from everything. The closest university, Southern Illinois University at Carbondale (SIUC), was ninety minutes away, and it was not a large research institution.

What was I to do? Keep my job in Indianapolis, and go back to work as soon as maternity pay expired? I did not make enough money to pay the mortgage, let alone all the added expenses for our new baby. My husband's salary was paying all our bills. Should he try to find a new job? Of course, but the right job is very difficult to find and would have required us to move again anyway. We took the transfer, but we decided to live between SIUC and his job just to give me potential career opportunities.

After moving, I stayed the course as a stay-at-home mom. In a neighboring town I found a group called La Leche League that offers support and help to breast-feeding moms and decided to attend the local meetings. I got to meet people with children the same age as mine. We formed a playgroup and became friends, and I mourned the time I had wasted working so hard at a career that had seemingly vanished. Why did I spend so much of my twenties working for a career in science? I wondered what part of my schooling was helping me teach my son how to use the potty. I knew that someday I would be able to teach my son about molecular biology—what a small consolation prize for eight years of exhausting work on top of my undergraduate degree.

Then, through La Leche League, I met another mother who was an artist-illustrator. During our conversations I learned she was working on a textbook with one of the professors at SIUC. Surprisingly, this was a molecular biology textbook, and she was doing all the illustrations. I mentioned that I had a PhD in that field and would love to be involved if she ever needed any help. About a year later, the professor did need help to finish the book, and I was hired to read and critique the chapters. He even offered to let me write one or two of the chapters. I was now actually using my degree. . . . what a trip! Of course, I was dumbfounded that he even hired me, especially since I was pregnant with my second child.

Writing was the best opportunity in the world: I worked during naptime, after the kids went to bed, and before they woke up. I took time off after the birth of my second son, and then we got news from the publisher that they wanted this one textbook to be split into two books, and they wanted me to be named as an author on the second volume. I went from being an editor-reviewer to an *author*? Well, books aren't completed overnight, and the second volume needed some introductory chapters to make it an independent textbook. I was the one to write these chapters, and I had also written two of the later chapters—quite the job for a stay-at-home mom. I felt at least partially vindicated. At least I was able to inspire and teach others about the mystery of science by writing! I was putting those years of graduate study to use!

Just recently, my husband's company decided to move us again. His latest transfer took us outside St. Louis, and the door remained open to my work on the textbook. With the project primarily accomplished through e-mail and phone, living a couple hours away from my coauthor has not been an issue. I have taken an adjunct teaching job at the community college, I have two wonderful boys and a daughter born early 2007, and the book is now in the final stages before going to press. I don't know what the future holds. I don't know where or when the next opportunity will appear. I want to continue to write, edit, and be involved with the book world. I love teaching part-time. I can envision my future with so many different permutations: I could pursue teaching full-time at the community college, where I could have every summer to spend with my children. And I could write many books and try to publish as much as possible, staying at home full-time. Or I could simply be a stay-at-home mom. These days I do not see a clear path, as I did when I entered graduate school. Being a scientist and a teacher are still my goals, but I see these roles so differently than I did when I began my PhD. I see how writing can influence a reader to pursue a degree in the sciences. I see that being involved in my son's elementary school can inspire a future scientist. I see that my life within La Leche League is an opportunity to teach about the biology of breast-feeding. I don't have to be confined to one university to be a successful teacher and scientist. The world around me is my own sphere of influence.

Variety, Challenge, and Flexibility

The Benefits of Straying from the Narrow Path

Marguerite Toscano

Quaternary Marine Geologist; Smithsonian Institution Research Associate,
Department of Paleobiology; Editor for the Association for Women
Geoscientists (AWG)

PhD, Quaternary Marine Geology, University of South Florida, 1996

Part-timing was never in my grand scheme. I was groomed for academia by my master's thesis advisor at the University of Delaware and went on to get a PhD, thinking and knowing this was my path. After looking into situations where both my husband and I could work and study in the same location, I chose my PhD program, in large part, because my husband was offered a good job nearby. Unfortunately for me, this turned out to be the wrong program, but I toughed it out for six grueling years because transferring would have meant a long-distance marriage and loss of my life outside school.

When I finished my degree in 1996, I went a year without a job and had my first and only child. I applied for jobs but had few viable options, and my dismal PhD experience left me with little enthusiasm for academia (my sole career objective). After my daughter was born, I applied for a couple of alternative jobs, but my heart wasn't in them either. In the meantime, a well-meaning older friend gave me a subscription to a publication aimed at stay-at-home mothers; she meant for this to help me be at peace with my situation. The articles were 100 percent in favor of staying at home, and rather than helping, they depressed me, calling my motivation for work and my own intellectual needs into question as contrary to my child's need for a full-time mother. The effect was profound. Though I had made the short list for two jobs, I was heavily impacted by a baby who never slept and the accompanying mental and physical fatigue. Then there was the magazine, forcing me to ask, What kind of mother was I? How could I leave my baby in a strange place with strange people while I worked? Needless to say, I was neither mentally engaged nor well prepared for the interviews and didn't get either position.

My husband transferred to Washington, DC, when our daughter was one year old. In DC I heard about a postdoctoral position on a topic only marginally related to my background, but I had read Peter Fiske's book *Put Your Science to Work* and was determined to try something, anything, that would use my broader knowledge, skills, and scientific talents. I felt it was an obligation, and I felt that reverting to a stay-at-home mom would stall my career and take me further out of the loop. Of course, this meant day care for my then outgoing eighteen-month-old, but luckily a good center was available right in my husband's office complex. For a couple of years it worked well to have a full-time day care situation in close proximity to one parent. The postdoc was more difficult. My sponsor was looking for a personal assistant, not a collaborator, and for three frustrating years I had to assert myself for my own research time and resources. Again I stuck with it despite my misgivings; my daughter was taken care of, and I was *supposed* to be working. I was tired, overworked, and cranky. Fortunately, at the same time I also held a collaborative appointment at the Smithsonian Institution with a scientist who was conducting research in areas of great interest to me. We established a good working relationship and collaborated on several papers.

When it came time to switch from day care to regular school, the logistics become even more difficult because of the 8:30 to 3:30 schedule. This is where things got interesting. My husband and I alternated drop-offs and pickups so each of us could get in a full day, which sounds easier to work out than it was. When I finally forced myself to leave the postdoc, things changed. I took on all school-related duties.

I continued in my appointment at the Smithsonian Institution, which allowed telecommuting and provided some flexibility around my daughter's school schedule. In all this time I haven't placed my daughter in after-school care, preferring that she spend that time at home with me doing her homework and decompressing from the school day. Sometimes she stays late for an extracurricular activity, which helps me get in a longer day downtown once or twice per week.

Currently, I still do unpaid research with my good and generous mentor, whom I truly like and respect; I am blessed with several excellent working colleagues and am well networked and respected in my field. Finding funding is a constant quest. I also wrote educational science articles for a new Smithsonian website. Contractual jobs like this help pay for my commute and other expenses and give me the flexibility I need. I can work and publish, schedule trips to the office around my daughter's schedule, and volunteer when needed. Additionally, I was the editor for the Association for Women Geoscientists (AWG) and for four years had the challenging job of creating AWG's bimonthly newsletter and promoting the organization's mission. I am also a part-time professional musician and member of the board of trustees of a major performing arts organization in DC. My life is very busy, interesting, and varied. I meet and work with admirable, high-caliber people from many professions in the course of my various "jobs." Given this variety, I am not intellectually limited—though it seems I can have professional satisfaction, recognition, and flexibility *without* a paycheck (or not much of one) but rarely *with* one. The idea of finding yet another full-time position rife with stress and lacking in professional and personal support keeps me from actively looking, although I would love to find the right one. Last, the thought of having no time for a personal life and the cultural activities that also define who I am makes the part-time situation sensible, at least for now.

The Balancing Act

Kim M. Fowler

Senior Research Engineer, Pacific Northwest National Laboratory

MS, Environmental Engineering, Washington State University, 1996

I am lucky. I have two healthy and smart sons, an amazing husband, supportive parents, and a career doing the research about which I am passionate. My "luck" is managed through the act of balancing my professional, family, and personal aspects of life. When life is running smoothly, the balancing seems natural and easy. However, there are days and weeks when one part of my life needs more attention than others. If I can't gain the balance I seek in a day, I look for it within a week or month. The daily imbalance doesn't cause the same level of anxiety that it used to. To manage the daily imbalance, I use the advice a doctor gave me about my young child's eating habits: it doesn't matter that he eats a perfectly balanced diet over a day or even a week, as long as over a month there is some resemblance to a healthy diet. Allowing for give-and-take on the key aspects of my life is how I attain my goal of balance. Coincidentally, my research is in the field of sustainable development, which involves the balance of environmental, economic, and social equity considerations.

I believe that balancing professional, family, and personal considerations involves working on the things you are passionate about, finding good people to work with—including role models and mentors—gaining technical expertise, and receiving support at home for who you are, your routines, and your work schedule. Finally, balancing involves losing the guilt about not being home twenty-four hours a day, seven days a week; finding regenerative activities (ideally ones that involve the family); and setting boundaries so that you are able to treasure the time you have with family, friends, and by yourself.

> To be successful, the first thing to do is fall in love with your work. (Sister Mary Lauretta, American nun and science teacher)

I define "passions" as the things you do or think about that give you positive energy. For example, one of my core passions is the drive to learn. I love learning new things. What I choose to learn is driven by my other passions. Being able to work on topics to which I have a strong personal connection makes the day more enjoyable. Of course, I have passions that do not correlate with my professional world and work tasks that don't match up with my passions. This is where the balancing act enters the picture. With a flexible schedule and work environment, I try to manage my work commitment in a way that minimizes the work that doesn't energize me and that allows for me to pursue my outside interests as well. On those days when I am not excited about my work, I go through a mental checklist to see if my interests or passions have changed (over time they have matured and the priorities have changed), and then I look at the tasks I have ahead of me. If there isn't a correlation, I figure out how to adapt my work so that it is "fun" again.

> Keep away from people who try to belittle your ambitions. (Mark Twain)

Early in my career I learned the value of working with supportive people. I am a high-energy person and tend to set goals and work hard until they are accomplished. I have high expectations for myself and others. As a young, eager researcher there were many peers who warned, "You are going to burn out!" Eventually, there were enough "burn out" statements that I started to doubt myself.

Because of these doubts, I started looking around for mid- to late-career professionals who still demonstrated a drive to accomplish meaningful goals. I found one right under my nose: my father. Funny how I hadn't noticed before what a unique professional he is. In what is now a more than forty-year career, he has taught high school mathematics and science and college-level mathematics and education courses, coached girls' and boys' sports and math competition teams, and held seminars to bring the metric system to teachers (back in the seventies). He is currently a leader in our state's science and mathematics education reform effort. All his work centers around his passion for improving science and mathematics education (yes, even the coaching has an element of that). He's passionate yet patient. He works hard. He has a vision and doggedly pursues it to its logical conclusion. Almost everything he touches succeeds, and he hasn't burned out yet.

What a relief. Not only did I have proof that everybody doesn't burn out, but I also felt that if my dad was able to keep his passion throughout his career, then I had a chance. I already had tremendous respect for my father, but now he is my professional role model. Although we work in different fields, we are both trying to effect change, and from him I have received considerable guidance on the general challenges I have encountered. Most important, finding a role model has allowed me to pursue my passions without concern for burning out.

> People who know little are usually great talkers, while people who know much, say little. (Jean-Jacques Rousseau)

When I selected sustainable design and development as my technical field, very few people knew what it meant. I was determined this field was the correct one for me, so I pursued it by reading everything available and then developing strategies for strengthening the science behind sustainable design. I connected existing, accepted fields, such as pollution prevention and energy efficiency, with the science needs of the developing sustainable design field. This resulted in small-scale research projects that eventually grew as the field became more accepted. I developed my external reputation through publications and software development. As I built up my technical expertise, I also needed to explain the field to others. I taught graduate-level courses to expand the understanding within the engineering disciplines. As my reputation outside the Laboratory grew, my reputation within the Laboratory grew as well. Now the Laboratory has adopted sustainable design as one of its facility design expectations.

> The antidote for fifty enemies is one friend. (Aristotle)

I have been fortunate to have the unquestioned support of my parents, husband, and a core set of friends. My parents have been supportive of my various interests throughout my life. They have encouraged me to pursue mathematics and science when there were few women in the programs. They continue to be supportive by helping my family when I travel and when an extra adult is needed to get the boys to their various venues. My husband has supported my pursuit of my dreams from the day we met. He appreciates the passion I have about my work and encourages me to pursue what makes me happy. My friends tend to be connected to other parts of my life—work or family. The balancing act with friends means that I rarely have time just to sit around with a friend, but rather those interactions are connected with other activities. For example, I have working lunches with my professional friends, and our family friends have children of the same age, so we combine "play dates" for the boys with adult friend time for me and my husband.

> One of the secrets of life is to make stepping stones out of stumbling blocks.
> (Jack Penn, in *Good Stuff,* edited by Ken Dooley)

During a regular workweek I try to leave the house before the boys wake up. Although I am a night person, I get to work early so that I can get a jump on my work and have a more flexible end-of-the-day work schedule. The boys know not to expect me at breakfast, and they know that dinner depends on my workload, but they can count on me for story time each night. With respect to business travel, we have other routines. Prior to a business trip I prepare notes for my children that are delivered while I am gone, and once I'm away I call around bedtime to discuss the day's events and wish them good night. My husband treats them to special dinners when I am gone, and my parents offer regular backup for my normal duties. I don't bring the kids gifts from each trip because I want them to be happy just to see me. Although my boys don't like it when I travel or work late, these routines and my support network make it more bearable for them.

> You must do the thing you think you cannot do. (Eleanor Roosevelt)

My schedule is, like many, overbooked, yet it is part of my core belief system to contribute to the community. My comfort area is to volunteer in my kids' schools but also includes coaching sports and academic teams and leading Cub Scout den activities. When I volunteer, I let the teachers know that I am interested in science and mathematics; I try to volunteer to work with the kids in the classroom and be as flexible as I can to meet the teacher's schedules and classroom needs. As part of its outreach program, my organization asks researchers to provide hands-on science lessons to the local schools. I participate in this program at work and have been able to be a scientist volunteer in my sons' classrooms. I realized how important these visits were when one of my son's tried to teach me something that I had taught him in the classroom. He remembered the lesson but had forgotten that I was the one who had given it.

> Cleaning your house while your kids are still growing is like shoveling the walk before it stops snowing. (Phyllis Diller)

I have spent immeasurable energy feeling guilty about not being everywhere and doing everything. Research tends to require long hours of work. I have learned that wasting energy on guilt doesn't help anyone. Guilt does not get my work done more quickly or allow me to spend more time with my family. There are only twenty-four hours in a day. My philosophy is to enjoy every minute, even the times when I am being challenged. While I have a high-energy personality, my husband is more relaxed and calm. Our children are different from each other and different from both of us. My husband's ability to get things accomplished in a calming way creates a positive atmosphere in our home. To reach this point we found that we had to let go of keeping the house spotless or in perfect order. As soon as we outsourced housecleaning, everyone was happier. We would gladly purchase other, similar services if it would reduce our stress level.

> Nothing great was ever achieved without enthusiasm. (Ralph Waldo Emerson)

The work that I do is all about balancing competing interests. I have applied these skills to my life in general and attempt to balance my work, personal, and family life on a daily basis. Hard work takes energy, but some work also gives back energy. I aim to work on things that give energy back to me. Here are three pieces of advice I have for other mothers pursuing technical research fields. First, follow your passion. Second, develop a strong technical reputation. And third, yet most important, love your family and enjoy the time you spend with them.

Juggling through Life's Transitions

Cal Baier-Anderson

Health Scientist, Environmental Defense

PhD, Toxicology, University of Maryland, 1999

I am an only child, and my mother was a stay-at-home mom. I have no doubt that this simple fact influenced my chosen path. My mother was a bright woman, but her personal growth was severely limited by her role at home. It didn't have to be that way, but it was, and this had its impact. Her pain spilled over into my life. In response, I concluded that my life would be different. There are two fundamental assumptions that form the foundation of my life: that I will have a career that will allow me to be financially and emotionally independent and that the work that I do will somehow contribute to society. There was never any doubt that I would manage kids and a career or that my husband would take equal responsibility for raising a family and making the household function. All other aspects of my life must mesh with these basic facts.

While my two children were young, I earned my doctorate degree in toxicology. My graduate advisor, married to a researcher at the National Cancer Institute, faced a similar situation, raising two boys while he and his wife juggled their careers twenty years earlier. As a result, he was tolerant of my family needs, and my lab schedule was manageable. After graduation, I continued on as a research associate at the University of Maryland, Baltimore, which evolved into a job as an assistant professor with a 51 percent non-tenure-track appointment. Though in any given week I worked between twenty-four and forty hours, my schedule was extremely flexible.

I attended faculty meetings and had cordial relationships with other faculty members, who seemed to accept my part-time status, but I was not successful in getting that "big grant." This definitely separated me from the other professors. Meanwhile, I became increasingly interested in understanding the effects of chemicals in the environment and on human health. I developed and taught a class in chemical risk assessment. I worked with communities living adjacent to hazardous waste sites, helping them to understand health risks and participate as stakeholders in the cleanup process. The small grants that funded my community work were not as highly valued as the big grants. So although at any point I had the option of converting to a tenure-track position, I found that option particularly daunting given the emphasis on building a grant portfolio.

To supplement my income, and to fill in downtime (because I like to keep busy), I worked part-time for a small consulting company that provides toxicological risk assessment services, and I took on an occasional interesting project as an independent consultant. Juggling two or more jobs still allowed me to work at home two or three days a week. My home office window faced the woods, so I could listen to the birds at the feeder and the wind chimes on the deck. I watched hundred-foot trees sway in the breeze and frequently thought how lucky I was to have this home and the flexibility to enjoy it. It was a good balance: I found the work intellectually rewarding and my schedule was very flexible, although I did often work seven days a week. My husband and I both managed to juggle doctor and dentist appointments, parent-teacher conferences, shopping, and the boys' hockey and lacrosse schedules.

But something was missing. In part because of my growing interest in environmental advocacy and the desire to contribute to change, combined with the realization that my research interests were not particularly fundable within the context of my academic position, I recently applied for— and accepted—a job with Environmental Defense in Washington, DC. Accepting the job meant giving up some flexibility, particularly since I was asked to continue teaching my risk assessment class in Baltimore. But I am impressed with Environmental Defense, and it provides me with sufficient flexibility to manage work and family. With both boys in high school, it seemed to be a good time to make my move.

Our home life is generally chaotic. Our house is on five acres; we have two sheep, two goats, and three dogs. Everyone pitches in, but it is hap-hazard and sporadic, and I have had to learn to ignore what I cannot (at the moment) change. Occasionally there are lapses—a missed appoint-ment, a mistaken game time. The dishes pile up; the dust bunnies tumble across the floor. The kitchen renovation, begun more than five years ago, is not yet complete; the walls are covered with the fingerprints of child-hood, and in nine years of living in this house we have never cleaned the windows. With a life this busy, you have to make choices, and there is no time to wallow in regret.

Juggling family and work seems a lot to me like weaving an elaborate blanket. There are different strands for work, husband, kids, school, sports, doctor appointments, shopping—so many strands! And each day I select the strands I need to work with, choosing what is most important. The tex-tures may not match and the colors may clash. There are probably dropped stitches and even knots of clumsy repair. But if you were to step back, the blanket would be rich and warm and, in its own way, beautiful.

Having It All, Just Not All at the Same Time

Andrea L. Kalfoglou

Assistant Professor, Health Administration and Policy Program, Department of Sociology and Anthropology, University of Maryland, Baltimore County

PhD, Health Policy and Bioethics, Johns Hopkins Bloomberg School of Public Health, 1999

When I began my doctoral program in 1994, my mother asked me how I planned to both raise a family and have a high-powered career. My response to her was, "It isn't the administrative assistants who have the power to negotiate their terms of employment." I also loved my field and couldn't imagine doing anything else but pursuing my doctorate. I've managed to have two children and an interesting career, but it has required major sacrifices and has been physically and emotionally demanding. Along the way, I've learned a few things.

You have to be willing to make sacrifices. My husband and I decided early on that our family would stay rooted in one community near the Washington, DC area. Our extended families were local, and we had a lot of financial stability through my husband's job. After I received my degree, a fabulous job opened up in San Francisco with my name written all over it. I didn't even apply. When colleagues I respected begged me to apply for faculty positions at their institutions, I had to smile and say, "Thanks so much for thinking of me, but I can't relocate." Instead, I've scrounged locally for work, changing jobs frequently when the soft money dried up, and waited for a faculty position to open up, knowing it was a long shot.

There have been pros and cons to this arrangement. My husband's job is more stable and essential for our financial well-being. When our first son was born, my husband's boss gave him a tongue lashing about taking two weeks off to care for me after my C-section. He wanted my husband to understand that his job needed to be his first priority. As a result, we protect his job usually at the expense of mine. Whenever a child is sick, I usually stay home. Even if it were easy for my husband to take time off to care for a sick child or attend school activities, my sons demand Mommy, and I want to be there for them. While my employers never said anything directly, I know I've been perceived as less reliable. On the positive side, there has been less pressure on me to be the financial provider for the family. This has made job security less important and has allowed me to try new things and escape bad work environments.

At times I felt like a pioneer "coming out" as a working mother—storing pumped breast milk in the community refrigerator, leaving early to pick up my kids after school, and occasionally bringing an infant to work with me. I believe that the workplace needs more working parents to put pressure on the system so that there can be positive changes, but as an individual working woman looking back, I think I've hurt my career by being too open about my situation. If I had advice to give to other working women thinking about starting a family, it would be to keep the workplace less personal. If you are taking time off work to care for a sick child, take your child to the doctor, or chaperone a field trip, telling your boss you have "an appointment" is sufficient.

Be open to alternative career paths and prepare to encounter many bumps in the road. When I became pregnant for the first time, I was working for a pseudoacademic organization. Employment was from contract to contract, and my first project was nearing completion. I received an offer for a new project with a promotion. The major problem was that my baby was due at the same time the project was due to be completed. My first project had required fourteen-hour days to complete on time and within budget, and it had left me physically exhausted. I turned down the promotion and instead accepted a tentative offer in a more supportive roll. I approached human resources (HR) about how to handle my situation. I was afraid that if the senior management knew about my pregnancy, the tentative offer would mystically disappear. My HR representative agreed that I should keep my pregnancy a secret as long as possible. I learned from her that the senior managers were all deeply concerned about my decision and wanted to know why I "lacked career ambition." Eventually I explained to my new boss why I had decided it was in everyone's best interest that I decline the promotion. Although everything worked out, and I was able to keep my job, I had to identify and speak with all the senior managers who believed I lacked ambition in order to set the record straight and protect my reputation.

I went into labor while at work and returned to my job after only seven weeks of maternity leave—following a C-section—because I was so concerned about my project faltering without me. I expected that this devotion to the organization and my project would be acknowledged. I was able to ease back in at 50 percent time. It made adjusting to first-time motherhood bearable. I was so exhausted most days that I would fall asleep at red lights on the way to work. In hindsight I don't believe my sacrifice was appreciated, and I wish that I had taken a full three months of maternity leave. You never get that time with your infant back.

By the time I became pregnant with my second son, I was working 80 percent time for a grant-funded center within a university. It had great potential for promotion into a faculty position, and I was throwing myself headfirst into my work. Because of the amount of stress I was under, I began taking prescription medication for insomnia, anxiety, and depression. On the day I discovered I was pregnant, I stopped taking all the medication. This led to a week of not eating and sleeping, which resulted in a full-fledged psychiatric crisis. I spent a week sitting in a chair at my parents' completely unable to do anything but cry. I remained clinically depressed and on medication that was potentially dangerous to my fetus throughout my pregnancy but managed to keep functioning—barely. Then my husband and I struggled through a period of time when test results indicated our son was at increased risk for Down syndrome and when sonograms showed a hole in his heart. (Thankfully, neither of those things manifested.) I let my boss know immediately about my pregnancy and psychiatric condition. She promised to be supportive but then gave me the worst performance review I have ever received. When the funding cycle ended, so did my job. I left feeling completely defeated. Millions of women manage this balancing act, but I couldn't hack it. I was prepared to drop out of the workplace completely and pull my sons out of day care.

Fortunately, I landed in a great fellowship at the National Human Genome Research Institute where I was again able to negotiate an 80 percent effort. It has literally been a career saver. My project responsibilities have been manageable. It has given me two years to recover emotionally and physically. And I've had time to think about what I want for my future. Because I've drawn certain lines in the sand regarding the amount of time I will spend working, I've again been pegged as a mommy-tracker, but at least I have had options.

Because I had the freedom to really explore my career options during my fellowship, I became aware of a number of new opportunities within the local universities. This coming fall, I will begin a tenure-track faculty position teaching undergraduate public health courses. It is hard money for a nine-month contract with the opportunity to buy out teaching time with research grant funds. On the basis of everything I know so far, it appears to be my dream job. The seven years of research, policy, and publication experience set me apart from the competition—many of whom were coming straight out of graduate school. With my childbearing years behind me and my depression and anxiety under control, I feel ready to take on this new challenge with enthusiasm.

Plan your child care before the baby arrives. For some reason, during my first pregnancy, I just assumed we would be able to find quality, affordable child care without too much effort. By searching the licensed in-home centers in my zip code, we found a wonderful retired teacher who had only one other infant in her home. We shared the same religious tradition, which created an instant bond. She loved our child as if he were her own grandchild. In retrospect, I now know how lucky we were to have found her. When planes tore into the Pentagon and when a sniper picked off people blocks from our home, I knew my son was safe in the basement of this woman's loving home.

But with time, her knees gave out, and it became increasingly difficult for her to care for children. At my first son's conception, we had him on a waiting list for the day care center at my husband's employer. My son was finally offered a spot when he turned two, and we had forty-eight hours to make a difficult decision (a two-and-a-half-year wait is fairly common for quality centers in our area). We both felt he was ready for more peer interaction, but this day care center was a big unknown. It was expensive and would require us to bring our son on our two-hour round-trip commute into the big, dangerous city each day. Again, we were lucky, and we loved this center with its educated, committed staff workers. When our second son was born, he was welcomed at three months of age because of a sibling preference policy.

The hell was in the daily commute. We were all in the car before 7:00

a.m. and typically didn't get home until 7:00 p.m. The baby frequently cried the entire trip, while the three-year-old sat in a trance in front of the in-car DVD player. So much for what I had hoped would be quality family time. When I changed jobs to the fellowship out in the suburbs, my husband began doing this commute with two children by himself. The daily drop-off and pickup plus the two hours in the car with whiny children left him exhausted and grumpy. As if that weren't enough, we also learned that the building that housed our children's day care center was on a list of five financial buildings targeted by terrorists. The day after the press release, I suddenly burst into tears. Something had to change.

I looked into the day care facility affiliated with my institution and was told that I could check back after a year, as there were over a thousand children on the waiting list. I looked into local day care centers and was appalled by what I found—understaffed centers with infants crying alone in corners, centers where infants were expected to sleep with two- to five-year-olds running and screaming next to the row of cribs, day care workers who all spoke different languages and couldn't communicate with one another, much less the parents, and broken playground equipment on microscopic playgrounds.

We eventually found a center near our home and muddled through spending $30,000 per year ($45,000 of my pretax income) to make sure our kids were receiving decent care. I wonder all the time how single parents make this work.

I also did some serious thinking about the challenges of finding and keeping quality child care. While there are multiple benefits to employer-based child care, one of the drawbacks is that you lose access to that care if you leave your job. I realized that if our day care had been linked to *my* employment when I lost my job, I might have been in a position where I had to turn down the fellowship for lack of child care.

Join or create a network of other working parents. Other working parents are a wealth of information and resources. They have been my inspiration and support. I've built a network of women around me and make a point of trying to have lunch with one of them at least once a month. I've empathized over the phone with a friend who had her confidence shaken by a silly comment from a boss, and they've done the same for me. You don't have to work for a large institution to find this kind of support. There are community-based parenting Listservs, and you can create your own support group with a group of colleagues and friends. These are some of the best ways to get the inside scoop on work-related policies for parents, quality child care, and ideas for how to balance work and family. When a friend was harassed by a coworker about pumping breast milk at work, her e-mail network of PhD mommy friends rallied her to complain to her superiors. This effort resulted in an institution-wide e-mail reinforcing the pro-breast-feeding office policy. I've also created mentoring relationships with younger women that are mutually beneficial. Once I paid the conference fee for a student and asked only that she hold my infant son while I gave my talk. Without the shared hotel room and covered expenses, she never would have attended this conference that was beneficial to her career development. I've also found some wonderful books that focus on how working mothers are actually balancing career and family rather than on the many barriers to success. Finally, I really identify with the articles in *Working Mother* magazine.

To stay in, drop out, or have it halfway. There are benefits to the mommy track when you have young children. For the past six years, I've managed the balancing act but have realized that you don't ever actually achieve balance; it's a constant process of readjusting. Balancing has required giving up many opportunities and embracing these sacrifices as choices I've been lucky enough to have been able to make. I look back over these years and see that, in spite of the setbacks, I have accomplished a lot in my career. I've kept my foot in the door while tending to a young family. Dropping out would have meant losing my network of contacts, falling behind in my knowledge base, and losing the salary increases that have come with every new position. Jumping in full force would have jeopardized my health and my family's well-being. I value the time I spend with my children and the time I get to spend exercising my brain and interacting with my peers. My mantra has been "You can have it all, just not at the same time." When I've been frustrated by opportunities I've had to pass up and the stigma associated with being part-time, I remind myself that I still have many years ahead of me to accomplish my career goals as my children need me less. I've also realized that underlying much of the negative feedback I've received is envy—envy of my ability to set boundaries and of the time I have to spend with my family.

SECTION IV

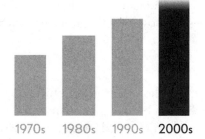

1970s   1980s   1990s **2000s**

WE ARE ALL IN THIS TOGETHER

The sun fades into the ocean. We lift a toast to the New Year and to the twelve consecutive New Years for which we have gathered from across the country. We are friends and colleagues, husbands and wives, parents and children. We are six individual families, and we are ten PhD scientists, two science teachers, one engineer, and nine children.

For twelve years we've set aside one glorious week a year, isolated from work, home, and extended family. We celebrate tenure awards and support one another through doubts about career choices and changes. While chopping vegetables for dinner or spreading peanut butter for the kids, we discuss the highs and lows of research, publications, and grants and the rewards and the difficulties of child rearing. We compare our girls with our boys, the sisters with the brothers, and we joke that only when our children become engaged to be married, or the equivalent, can they invite another body into our sacred space.

We celebrate our friendship and our relative good health by making fools of ourselves at the public basketball court—a little less agile than we were over twelve years ago as childless students, postdocs, and spouses, blowing off steam while playing ball at the playground back in North Carolina. We are a microcosm of women in science finding ways to make it work in the new millennium: three tenured university professors, one tenured community college faculty member, one stellar science teacher, and me, who still doesn't know how to categorize my life in science.

· · ·

The new millennium, like the seventies, is a time of change. Articles published in the *New York Times* have begun to question the ability and desire of women to combine the traditional model of success, and MIT reports that after reaching a peak in the 1990s the number of female faculty in most science departments has dropped.[1] Journals, conferences, Listservs, and seminars buzz, some in response to Harvard President Lawrence Summers's speech at the National Bureau of Economic Research Conference on Diversifying the Science and Engineering Workplace and some in response to the report *Beyond Bias and Barriers: Fulfilling the Potential of Women in Academic Sciences and Engineering.*[2]

Summers's attempt to deliver a provocative rather than an institutional speech—hypothesizing about the underlying causes for the paucity of tenured women science and engineering faculty at the more prestigious institutions—more than succeeded. Several excerpts are provided below:

> Twenty or twenty-five years ago, we started to see very substantial increases in the number of women who were in graduate school in this field. [Summers broadens the issue beyond just science and engineering and refers to professionals in health, law, business, and other professional services.] . . . If you look at the top cohort in our activity, it is not only nothing like fifty-fifty, it is nothing like what we thought it was when we started having a third of the women, a third of the law school class being female, twenty or twenty-

1. Andrew Lawler, "Women in Science: Progress on Hiring Women Science Faculty Members Stalls at MIT," *Science* 312 (2006): 347–48; Louise Story, "Many Women at Elite Colleges Set Path to Motherhood," *New York Times*, September 20, 2005; Eduardo Porter, "Stretched to the Limit, Women Stall March to Work," *New York Times*, March 2, 2006.

2. National Academy of Sciences, National Academy of Engineering, Institute of Medicine of the National Academies, *Beyond Bias and Barriers: Fulfilling the Potential of Women in Academic Sciences and Engineering* (National Academies Press: Washington, DC, 2006).

five years ago. And the relatively few women who are in the highest ranking places are disproportionately either unmarried or without children, with the emphasis differing depending on just who you talk to. And that is a reality that is present and that one has exactly the same conversation in almost any high-powered profession. What does one make of that? . . . [T]here are many professions and many activities, and the most prestigious activities in our society expect of people who are going to rise to leadership positions in their forties near total commitments to their work. They expect a large number of hours in the office, they expect a flexibility of schedules to respond to contingency, they expect a continuity of effort through the life cycle, and they expect—and this is harder to measure—but they expect that the mind is always working on the problems that are in the job, even when the job is not taking place. It is a fact about our society that that is a level of commitment that a much higher fraction of married men have been historically prepared to make than of married women.

It does appear that on many, many different human attributes—height, weight, propensity for criminality, overall IQ, mathematical ability, scientific ability—there is relatively clear evidence that whatever the difference in means—which can be debated—there is a difference in the standard deviation, and variability of a male and a female population. And that is true with respect to attributes that are and are not plausibly, culturally determined. . . .

[M]y sense is that the unfortunate truth—I would far prefer to believe something else, because it would be easier to address what is surely a serious social problem if something else were true—is that the combination of the high-powered job hypothesis and the differing variances probably explains a fair amount of this problem ["this problem" referring to the underrepresentation of tenured women faculty in science and engineering at prestigious colleges and universities.][3]

Those observations offered by the president of one of the top institutions of learning in America precipitated important discussions and attention to the challenges facing women in science.

Not more than a year later, the release of *Beyond Bias and Barriers*—

3. Lawrence Summers, "Remarks at NBER Conference on Diversifying the Science and Engineering Workforce," Cambridge, MA, January 14, 2005, http://www.president.harvard.edu/speeches/2005/nber.html.

though confirming the inconsistent relationship between women receiving PhDs in science and those attaining "top positions," in this case full professorships—provided a suitable and unplanned follow-up to his comments.

> Women have the ability and drive to succeed in science and engineering. Studies of brain structure and function, of hormonal modulation of performance, of human cognitive development, and of human evolution have not found any significant biological differences between men and women in performing science and mathematics that can account for the lower representation of women in academic faculty and scientific leadership positions in these fields. The drive and motivation of women scientists and engineers is demonstrated by those women who persist in academic careers despite barriers that disproportionately disadvantage them.
>
> The evidence demonstrates that anyone lacking the work and family support traditionally provided by a "wife" is at a serious disadvantage in academe. However, the majority of faculty no longer have such support. About 90% of the spouses of women science and engineering faculty are employed full-time; close to half the spouses of male faculty also work full-time.
>
> The fact that women are capable of contributing to the nation's scientific and engineering enterprise but are impeded in doing so because of gender and racial/ethnic bias and outmoded "rules" governing academic success is deeply troubling and embarrassing. It is also a call to action."[4]

In the first decade of the new millennium, women will represent more than 30 percent of all science and engineering PhD recipients. There is no doubt that the door to science is opened wider for these women, and in many cases the environment is more inviting than for those who stepped across the threshold in the 1970s. Even so, one may be surprised by essays in this section that sound as if they could have been written decades ago. Here recent graduates and graduate students add their voices to those of the scientist-mothers we heard from in previous sections. This is their call to action.

4. *Beyond Bias and Barriers*, 2–12.

Exploring Less-Traveled Paths

Deborah Duffy

Center for the Interaction of Animals and Society, School of Veterinary Medicine of the University of Pennsylvania

PhD, Psychology, Johns Hopkins University, 2001

When my daughter was born, I was on the traditional academic path working as a postdoctoral fellow (studying animal behavior) and applying for tenure-track jobs. I had all sorts of wonderfully naïve fantasies about how I would seamlessly blend being a mother and an academic. I pictured myself serenely breast-feeding my daughter while working on a manuscript or carrying her in a sling while attending meetings and seminars. Boy, was I wrong! Breast-feeding didn't come easily, and there was nothing serene about it for the first few months. Usually, by the time I managed to get her into the sling such that I was reasonably confident that I wasn't going to smother her, it was time to feed her . . . again. No, this blending of science and motherhood was not what I had envisioned at all.

My maternity leave was supposed to last only six weeks. It ended up lasting three months. My postdoctoral mentor, bless her heart, was understanding and did not pressure me about returning to work. Though she didn't have children of her own, and didn't know much about what I was going through, as a biologist she was fascinated and asked lots of questions about the experience. She was, and still is, very supportive of my research and career choices, even those that are somewhat less than traditional in academia.

Going back to work was tough. I dreaded dropping our daughter off at day care every morning. The university where I had my postdoctoral position didn't have on-campus child care, though there was a parents' cooperative that was run entirely by the parents who participated. I didn't know about the parents' co-op until after our daughter was born, and by then it was too late to join for that year. The other high-quality child care options in the area cost far more than what we could afford, so we had to settle for mediocre care, which emotionally made going back to work more difficult. Since then we have found wonderful and affordable child care, but I still suffer from a tremendous amount of guilt. I feel guilty that I'm apart from my daughter all day, and I wonder how my choice to continue working is affecting her. Then I remind myself that in addition to being her mother, I am also the most important role model she will ever have. I want her to know that with hard work and perseverance she can achieve her most ambitious goals. I want her to see that women can contribute to society in a number of ways, both inside and outside the home. And because I'm a scientist, I'm delighted to be an example of how girls really can excel in math and science! I also remind myself that there isn't really much of a choice for our family at all. We need two incomes to make ends meet, and we need my health insurance benefits, which are much better than what my husband's employer provides.

My husband works as an automotive technician (i.e., car mechanic) and has little flexibility in his working hours, which are long and tiresome. Most of his coworkers and supervisors have more traditional families in which Dad brings home the bacon and Mom takes care of the kids. If their wives do work outside the home, they still assume primary responsibility for all things home-related. Once, when our daughter became ill and needed to visit the pediatrician, my husband stayed home because I had a meeting that couldn't be postponed. When he explained to his boss that he wouldn't be at work and why, his boss asked, "Where's your wife?" I have often considered how balancing a career and a family would be a lot easier if I had my own "wife." Because of the lack of flexibility with my husband's work schedule, and the fact that I can do some of my work from home, I usually stay home when our daughter is sick. I occasionally have to remind my husband, and myself, that although the hours I work are somewhat flexible, the amount of work I need to get done is not.

While on maternity leave, I landed an interview for a faculty position. The phone interview took place from my home, and the call came just before my sweet baby was due to wake up, hungry, from her nap. Sure enough, about one minute into the interview she woke up and started fussing. My daughter was only about two months old at the time and I was still breast-feeding exclusively. There was no way that I, still clumsy at nursing, could handle the phone call and breast-feed simultaneously—that was a skill that would come weeks later. So I pacified her by letting her suck on my finger while I continued to talk with the very renowned professor from the very prestigious Ivy League university, trying to sound as coherent as a chronically sleep-deprived person can. Things got tricky when he started giving me pertinent information regarding my in-person interview in just three weeks (!!!). I rather impressed myself with my ability to keep one finger in my baby's mouth, cradle the phone under my chin, and write almost legibly. Things were getting closer to what I had envisioned during my naïve prenatal days, but the "serene" part hadn't yet materialized—and it wasn't likely to anytime soon with a job interview for which to prepare.

Before the interview, I had requested that the search committee provide a few fifteen-minute breaks in my schedule so I could pump breast milk. I knew a number of women who had breast-fed their babies, and one common fear was the dreaded wet shirt from leaking milk. Well, as interviews often go, things got behind schedule, and the first "appointment" to get knocked off the schedule was my pumping break. I was ushered from office to office and from building to building—and all along I carried my trusty breast pump! Fortunately, it was one of those professional-looking ones that could pass for a boxy tote bag. But as hour three turned into hours four and five, I could feel my breasts filling. I felt as if I were carrying a couple of overfilled water balloons on my chest. When I finally got the chance to pump, I was so tense from the interview process that the milk didn't begin to flow until my fifteen minutes were nearly up. As I sat there doing my best impression of a dairy cow, I recall thinking, "What a bizarre experience." I also thought about how unlikely it was that the other candidates found themselves in the same situation. I was just grateful that the co-chair of the search committee was a woman and a mother who had breast-fed her own children years earlier. Asking a man for a break to pump milk would have been very awkward. As natural as breast-feeding is, there was just something about drawing attention to my breasts during an interview that felt unprofessional.

I didn't get the job offer. The person they hired was a colleague and acquaintance of mine working in a very similar area of research. If truth be told, I was relieved. Though the location was great—less than an hour from where our families lived—I wasn't ready to balance motherhood and a tenure-track job in a department where the likelihood of getting tenure was about fifty percent. As it turned out, about a year and a half later I found myself working as a postdoc in the lab of the person they hired. So my family was able to live close to our relatives after all. Before that I did some part-time college-level teaching, which offered part-time pay but full-time work with lectures and lab exercises to prepare, exams to write, and papers to grade. My income didn't even cover child care expenses. I really enjoyed teaching, but we couldn't make ends meet with my low wages, so I called my newly hired colleague to ask if I could join his lab. I stayed there for one year while figuring out what my next move would be once my funding period ended.

My next move was to a whole new area of research. I had been considering for some time making a switch to work that was more directly applicable to daily life. I contacted the director of the behavior group at the University of Pennsylvania's School of Veterinary Medicine to see if I could do some informational interviews. As luck would have it, the director was conducting a job search for a research associate and asked if I would be interested.

I'm now working as a research associate studying applied animal behavior and find it rewarding because I can see how my work offers immediate benefits to society. The hours are great and flexible, and because I'm not managing my own lab, there is relatively little work-related stress. The salary is reasonable (of course, I still wish I were making more), and I have very good benefits. As with any research, finding funding sources is always a stressful issue. If we don't procure more funds soon, I'll be out of a job. I doubt I will remain much longer than two more years in my current position because there really isn't any room for advancement. My childhood dream was to work in a zoological setting, and while participating in a special one-day educational program at the Philadelphia Zoo, I met a helpful contact in the education department. We discussed her career path and the various types of jobs in the zoo's education department, where I now work as a volunteer two days per month to find out whether a career in wildlife conservation education would be a good fit for me. I'm also doing a bit of introspection to identify the skills that I developed during my training as a scientist that I most enjoy using and exploring careers that would draw on my artistic abilities as well.

Becoming a mother and trying to find a balance between career and family has prompted me to consider unorthodox ways of putting my scientific background to work. My horizons have opened up, and I feel free to explore career options that are more intrinsically rewarding to me. Most of my friends have tenure-track positions, and occasionally I feel like a failure when I compare myself with them. Sometimes they are envious of my normal work schedule and refer to my having "left academia," but that's not really true. I still do research at an academic institution, though I am no longer afraid of leaving the ivory tower. I still attend and present research at scientific meetings. I still prepare and publish manuscripts. I've simply opted out of the traditional tenure-track trajectory. In other words, I'm still in the neighborhood but I've exited off the main thoroughfare and am now taking the scenic route, which may or may not lead me out of academia. The scary part is, I have no road map, and I'm not quite sure where I'm going. That's another lesson I hope to teach my daughter by example: never let fear of the unknown stop you from exploring paths less traveled.

Standing Up

Gina D. Wesley-Hunt

Paleontologist and Evolutionary Biologist

PhD, Evolutionary Biology, University of Chicago, 2003

I was fired for getting pregnant, and it felt like a kick to the gut. I never saw it coming—who would? I was a postdoctoral researcher at a government institution, I was funded under a government grant, it was 2006, and I had all the pieces to build a brilliant career—or so I thought. I never expected my gender and reproductive status to destroy the path I was on. And make no mistake: this is a form of gender discrimination. The emotional scars and the career upheaval are still very fresh, but after giving birth to my daughter a few weeks ago, I am confident that I chose the right path. I stood up. I demanded that the institution change its family leave and anti-discrimination policy and the way it implemented the policy to ensure that what happened to me would never happen to others. I put my career on the line by cutting my postdoc short and risking professional censure. The final consequences are still unknown. Whatever the future holds, I will never regret making it clear to my advisors that their actions were discriminatory. Women before me who had demanded equality had given me the chance to follow a dream and a passion. I was not going to drop the ball. At this moment I am optimistic that mine will be a success story: the institution is making changes; I have won more than I lost by working within the system (although with outside help); and it looks as though future postdocs will be given the respect I should have received. On a broader scale, I feel that I have nudged the traditionally male-oriented scientific community one step forward and that maybe by the time my daughter has embarked on her career path, she will be able to be a mother *and* pursue her passions while living in a society that supports her choices.

When I was told that my two-year postdoctoral research position would be terminated a year early because of my pregnancy, I faced two paths. The first path was the safest for my career. I could accept my advisors' decision and leave quietly, but I would have to scramble for funding and another postdoc. The second path was much riskier. I could stand up for what I knew was right and put a spotlight on the discrimination that had occurred. This would mean that I could be considered a whistleblower and labeled a troublemaker, which could have significant impact on the possibility of being hired in academia. There would be no glowing recommendation from my advisors. Even if my advisors and other scientists at the institution were ordered not to retaliate, there is no real way to prevent the subtler forms retaliation can take. This is one of the big problems in academia. In many ways it is still a feudal system. Graduate students and postdoctoral researchers are beholden to their advisors for good recommendations and opportunities that can make or break their careers. I was completely at the mercy of the principal investigator (PI), the scientist awarded the grant funding my project. When I investigated what my rights were regarding family and sick leave, three offices at the institution told me it was at the discretion of the PI to offer family leave or terminate my position.

A postdoctoral position is generally a good time during an academic career to have children, as it is usually very flexible. It was my second postdoc, and my husband and I decided that now was a good time to start our family. I knew it would be a challenge, and I was under no illusion that having a child and continuing my career would be easy. But we were in an excellent position to make it work. My husband was willing to be the primary caregiver of our child so I could pursue my career. We had a stable home and were able to afford day care. So what could go wrong? If I couldn't make it work, I thought, who could?

Near the end of my first trimester in the spring of 2006, I was planning to inform my advisors of my pregnancy when panic struck. Another postdoc who received funding under the same grant was also pregnant. She was due in a month, and during a meeting with the PI, he let her know that after she gave birth, she would be terminated from the postdoctoral position. To prepare for my meeting with the PI and my immediate supervisor,

I investigated my rights at the institution. The equal opportunity office and the office overseeing interns and postdocs told me there was no policy that protected me. It was entirely up to my PI, and I was on my own. I voiced my incredulity at this response but was given no support. So I was at the mercy of one who could be considered my "feudal lord"—one who had ultimate power over my research career. But I pushed panic to the corner of my brain and rationalized that it wouldn't happen to me. The other postdoc had extenuating circumstances; without flexibility from the PI she could not work on-site after the baby was born. I did not have any such restrictions; I would be OK. So although I was completely without institutional support, I really thought that everything would work out. In fact, when I walked into the PI's office with my immediate supervisor, I thought we would be discussing how to approach my maternity leave, set research priorities, and so on. What occurred was not a discussion at all—they informed me that my position would be terminated at the end of the summer, one month before my child was due. In all fairness, my immediate supervisor offered to look into departmental resources to fund a position (though, not surprisingly, no funds were available) or to write a National Science Foundation grant with me; this would have to happen quickly, however, as the deadline was close, and then even if I were awarded the grant—with odds at about one in ten—I would still be without a position and funding for about a year. Regardless of these possibilities, I was being removed from my position because of pregnancy, with the consequence of having to find alternative funding.

After the meeting I was in disbelief. Then I was furious. I had known upon entering a male-dominated field that there might be some hurdles to overcome in my career. I had dealt with minor uncomfortable moments along the way, but never did I expect to deal with such blatant discrimination.

As my immediate response started to fade, I realized that I was in an ideal position to make some serious changes at the institution. The PI was not in my field or department. In my department I was surrounded by supportive colleagues: both senior scientists and younger cohorts. Therefore, I was much more protected from the ill will of my PI than most postdocs, and the consequences of standing up would be mitigated. So I started to speak out. I set up meetings and talked to many people and offices at the institution. I was not fighting for my position—I did not feel I could continue to work with my advisors—I was fighting to change the policy of the institution, and I made this very clear. It was a difficult time, and often I was saddened by the attitudes of those in power. During one meeting, the administrator in charge of all research and academics at the institution agreed that what had happened was unfortunate but that the PI was within his rights under the institution's policies. He said I was naïve to think I could change the institution. He offered no support and made clear his office would spend no energy to advocate for change.

As I continued to meet with administrators and scientists and made my situation known, the institution's lawyers got wind of what had occurred. One month after I was told that my position would be terminated, the lawyers forced the PI and my immediate supervisor to reinstate my position, with twelve weeks' maternity leave (unpaid—all I had wanted). At first, when reinstatement was offered, I was hopeful, thinking that maybe I could make this work. But then my supervisor wanted to discuss how I could make the PI feel happier about my continuation in the position. I pointed out that I expected the reverse. This suggestion was not taken well. After this conversation, any glimmer of hope died. I realized I would be under constant scrutiny, I would have to prove my worth daily, and it would be a horrible working environment. I decided to resign.

While all this was occurring, a second case of serendipity unfolded. A contact introduced me to a lawyer at the National Women's Law Center who offered me pro bono representation. She secured the services of a top law firm in the area in case the situation got sticky and we needed more resources. I now had a legal team working with me to convince the institution to change its family leave policies for postdocs. I knew this opportunity was unique and that I had to take it as far as I could to advocate for change. My husband and I had long conversations about whether we were willing to risk my future career. Would his job at the same institution be put in jeopardy? We decided to risk it all.

Although it was very tempting to jump straight to litigation, I wanted to stick to the high road and try to encourage the institution to do the right thing internally. Through my legal team, I made it clear that I would pursue external avenues if the institution did not take the steps necessary to prevent pregnancy discrimination from occurring again. In using these methods I weakened the threat to sue, but I could more easily convince people inside the institution to help pressure the administration to change its policies. Most people, scientists and office administrators alike, knew what had happened was wrong, and by working within the system I lowered the risk of being turned into the enemy. That said, it was still very difficult for me to inform the institution that I had legal representation. I knew I had crossed a line, that there was no going back, and that I had put my future in academia at risk. My advisors and office administrators no longer talked freely, and my visible involvement in the case decreased because everything had to go through the lawyers. But I know I received a lot more attention when I had the letterhead of two law offices backing me up.

We have traded letters with the institution twice now. They have not responded to the second. But I hear that changes are being made. It even appears that the changes will take place at a much higher level than I expected. In addition, a professor at another institution has used my situation to advocate for better policies at her university. I am optimistic that mine will be a success story. The momentum is building in the right direction, and I believe I will not have to use external forces to convince the institution to treat women equally. My fight, and the fight for family-friendly work environments in general, reaches beyond women—it affects fathers as well. Men should have every right to take paternity leave without damaging their careers.

My professional future is still bright, although I have taken a detour and the direction of my career has shifted. Most important, I have redefined what I consider success. Part of this new definition is standing up for rights I thought I had all along. Losing half my family's income was only part of the cost of fighting, but the reward is in helping women who come after me and in the hope that my story will convince other women to stand up rather than leave quietly to preserve their careers. I understand it is frightening to endanger one's career path and possibly abandon the hard work it took to reach this point in a career. If we don't stand up, discrimination will continue, and equality will take much longer to achieve.

We have to stand up and show society and our daughters and sons that mothers can be scientists too.[1]

---

1. Just before this manuscript went to the publisher the author sent the following note: I am happy to report that the institution has adopted a new nondiscrimination policy. Pregnancy discrimination is now highlighted as a form of sex discrimination (to avoid any confusion). In addition, anyone associated with the institution (not just employees) is now covered by the new policy and may utilize the equal opportunity office. However, there is no mention of granting the rights contained in the Family and Medical Leave Act to postdocs—there is still more work to be done!

Because of Our Mom, a True Rocket Scientist

Elizabeth Douglass

Graduate Student, Scripps Institute of Oceanography

Katherine Douglass

Physician

MD, Georgetown University, 2002

Elizabeth

I grew up as the daughter of scientists. Dad was a chemist, Mom was an atmospheric scientist, and I was one of five kids who found this arrangement completely normal. Both my parents had jobs, and the same was true for all of my friends' parents. As far as I could tell, "scientist" was just another job, like "teacher" or "firefighter" or anything else.

I'm currently a graduate student in oceanography, and it is only now that I really realize that science is different. Science is not an eight-hours-a-day, five-days-a-week job. Long hours, including weekends, are the norm rather than the exception. In addition, travel is an integral part of the process. As a field oceanographer I have been to sea twice, each time for more than two weeks. Additionally, attending scientific meetings is essential for presenting your work, meeting people, and making connections. I am single and childless, and the chance to explore the world through fieldwork and meetings is one of my favorite parts of being a scientist. However, the time required for advancement seems to make it almost impossible to have a career in science and a family. But people do it. Having my mom as a role model throughout my life has proved to me that it is possible, and the more people do it, the more accommodations are made within the field to allow for family life, and the easier it becomes. Maybe not easy, but definitely possible.

The main requirement for successfully raising a family and having a scientific career, it seems to me, is a deep love of science. I have always been aware of and somewhat amazed by my mom's passion for her work. Even as a kid, I remember her showing me pictures of model output that I didn't understand, talking about ozone and chlorofluorocarbons at dinner. The take-home message, which she communicated both verbally and by example, is that you'll never make it as a scientist unless you have what she calls "joie de science"—a pure love for science for its own sake, for its ability to enhance our understanding of the world we live in. I think she'd probably say that if you want to be a scientist, family or not, you won't succeed in this field if you don't love it; but if you do love it, that love will sustain you through the rough parts.

There are certainly other things that I learned by watching my mom. Most of her friends at work were men, and she did her best to be "one of the guys." Social interaction is important, and to succeed you need a measure of self-confidence and the belief that your work and ideas are important, along with a measure of willingness to listen to other people's ideas and integrate them into your own. I grew up to be like my mother, who tends to listen more than speak. She pushed me to be more outgoing and self-confident, knowing that this skill was important regardless of career choice.

I don't know whether I will have kids, but that's a personal decision rather than one dictated by my career. I know that I can both have a family and be a scientist. Mom did it, and there are several women at Scripps doing the same. Scripps is a great place to be a female oceanographer now: of the twenty-one students in physical oceanography, eleven of us are women. There are several women postdocs, and at least two of our female tenured professors are raising young children. Role models abound. And the field itself is becoming more accommodating—for example, child care is available at most major meetings. It still isn't easy. Balancing the requirements of fieldwork with family is hard; my mom gave up laboratory work and shifted to remotely sensed data in order to help keep the balance.

There are still advances to be made in attitudes toward the idea of combining a career in research with raising a family and in accommodations for those trying to do just that. Even today I have friends my age who have been asked, in the course of a job interview, if they intend to have children. But young scientists now have many role models like my mom and many examples of how to make life as a scientist and mother work.

Katherine

As a child, I never really thought things should be any different. My mom worked, my dad worked, and we all split up the chores. There were never any rules such as only Mom cooks, or Mom does the cleaning, nor did my two older brothers have specific "boy" chores while my sisters and I had different "girl" chores. We shared the jobs and responsibilities equally among everyone. We grew up exploring different options for child care, different babysitters, and some after-school day care solutions. On occasion, I remember wanting to come home right after school and not wanting to go to after-school care, but for the most part I thought it was just part of how our particular family was organized. In that framework, even the occasional feelings of resentment were not directed at either Mom or Dad specifically but at the two of them together as equals. Even as a child, I understood that they were at work and could not leave early. And through it all, I always felt as if I could do whatever I wanted to do with my life, that following their example, I would never be limited by being a woman. I thought both my mom and my dad were absolutely awesome and brilliant. My mom would talk about the ozone, and my dad would combine simple substances to make volcanoes: she's an atmospheric scientist and he's a chemist. It was partly my nature to consider myself invincible, but certainly nurture and environment have allowed and encouraged me to get where I am today.

It was in medical school, when an elderly surgeon made a point of introducing himself to male students only, even in a mixed group, that I started to realize some of the challenges that would lie before me as a female physician. Despite the presently changing gender profile of physicians, a lot of the behavior and tradition still come from the "old school."

Now sometimes I wonder: is it because of my youthful appearance or just because I am a woman that patients routinely ask me for the bedpan? I have had the experience of going through a complete introduction, medical history, physical examination (including a pelvic exam!), data collection and explanation, and discharge instructions for a patient in the emergency department, all the while wearing my white coat that clearly indicated MD, and then having the patient kindly and innocently say, "Yes, I'm feeling better and I'm ready to go home, but I thought I was going to see a doctor today." At that time, I sweetly smiled and said, "Ma'am, I am the doctor. Do you have any more questions?" Our encounter ended as she offered a slightly uncomfortable apology. I still can't help but wonder, who did the woman think I was, performing her pelvic examination? Likewise, I have had the experience of a multiply injured gunshot victim coming into the ER and, after I told him what to do, receiving the response, "Honey, who do you think you're talking to?" My male friend and coworker retorted more quickly than I could, "She's not your honey, she's your doctor, and you'd better listen to her."

By the time I finished medical school, my mother and I had had many chances to discuss the challenges of being a woman in science, particularly a successful woman in science. Going the extra mile is often required for women to achieve the same outcomes as men because of conscious and unconscious barriers. We talked about issues such as the delicate balance of personal expression, authority, and success, and the fine line separating an outfit that is feminine and professional from one that is revealing. We discussed the importance of utilizing personal, feminine strengths when entering leadership roles and the importance of using communication and collaboration to enhance professional relationships—skills that we as women bring to the table. We laughed about the benefit of entering a male-dominated profession because of short bathroom lines at professional meetings and the perpetual availability of a dance partner. And we have discussed the challenges of balancing career and family.

For my residency, I was very lucky to land at the Medical College of Pennsylvania (MCP), formerly known as Women's Medical College in Philadelphia. It is a place defined by its fascinating history as the first medical school for women in the United States. Here, as you walk through the main hall of administrative offices and conference rooms, ancient portraits lining the wall catch your eye; and as you look up at those portraits, you recognize that something looks different, that here in the hall of leaders, the white-haired individuals depicted are women, wearing dresses and long hair. These portraits set a tone for a truly enriching and inspiring place without so many of the gender stereotypes that continue to exist at many traditional medical schools. For the most part my colleagues and teachers at MCP made no distinction between men and women.

Upon entering the field of emergency medicine I learned that there are different challenges for female physicians. One is the persistent perception of male bosses that young women will soon need maternity leave, which they perceive as a disincentive to hiring a woman. Another is the perception among some men that the only reason a woman would choose the field of emergency medicine is the potential to pursue a part-time career and have a family. Many women, because of the requirements and expectations of medical school and residency, wait to have children until later in life, as their career paths can clearly conflict with starting a family. This should not, however, be a dissuading factor to pursuing a career path nor cause difficulty in finding a job.

I have been extremely lucky to find exceptional role models in both men and women in medicine, although for me there is something invaluable and particularly important about good female role models. During my residency, I had the good fortune to meet a friend and mentor in Sharon, a woman about ten years ahead of me in the career of emergency medicine. She has achieved a terrific balance in clinical work, administrative responsibilities, teaching and mentoring, and raising two beautiful twin daughters. She reminds me in many ways of my mom in her strength, knowledge, authority, and responsibility, as well as her compassion and femininity, which balance the equation. Sharon has taught me many skills helpful for managing gender-based perceptions and issues, along with her personal solutions for continued work in medicine while raising her children. Most important, she also taught me a lot about how to really be a good doctor through it all. As with my mom, we talked about the typically feminine traits of collaboration and communication and how these qualities are extremely effective and desirable and work to create even better emergency physicians.

For me it has also been a challenge to find a partner comfortable with my profession. In medical school in particular, I spent time socially with a group of mainly male medical students and a few women. When out socializing, we would witness these men meeting women (not in medicine), telling them that they were going to be doctors, and having women basically fall at their feet. We females had a slightly different experience. If we met a man and told him "I'm going to be a doctor," he would exit the conversation and often the room as quickly as possible. I now find myself with a wonderful, caring, supportive partner who is in no way threatened by my successes. He is supportive of my erratic schedule and my continued academic pursuits, and we have a strong respect and admiration for each other. He is a teacher. With all the choices that this generation of professional women has, it is not unusual for us to wait to start a family until we are established in our professions. This is quite a different path from that of my mother, who went through many of her professional development years as we all were growing up. However, I can assure you that as I face the challenges ahead in creating a balanced professional and personal life, and as I hope and plan to add children to that equation, I will look back on my mom and dad and the enabling, unbiased environment we grew up in and try as much as possible to emulate that.

Today I can say with certainty that my mom played a critical role in enabling me to become a successful woman in my profession. My dad did as well, and together they created an environment in which they taught by example that anything is possible. I think it is not too hard to believe you can do anything in the world when your mom is truly a rocket scientist.

On Being What You Love

Rachel Obbard

Postdoctoral Scientist at the British Antarctic Survey

PhD, Engineering, Dartmouth College, 2006

I am a scientist. That may seem simplistic, and in retrospect it has always been obvious, but it has certainly taken me a while to come around to it. I am also a mother. That, too, is obvious, although I like to think it's not unless the kids are standing right next to me. Like most working women, I worry that I am not getting the balance right and that I am not giving my children enough of my time and energy. But having struggled through single parenthood and a midlife career change to become someone I am finally proud of, I have come to the conclusion that the single greatest thing I can convey to them is the conviction to follow their dreams.

Looking back, it has always been clear that I am a scientist. It is reflected in my interests and in the way I approach things. As a child in the 1960s, I had a little sister, Harriet, who had the biggest collection of Barbie dolls and—thanks to a munificent second cousin—the largest selection of Barbie clothes that I have ever seen. I had one doll, Barbie's teenage sister, "Skipper," who always wore the same T-shirt and overalls. Harriet frequently asked me to "play Barbies." How did one do such a thing, I wondered? It seemed to involve long conversations and many changes of clothes, neither of which appealed to me. So when asked, I would offer instead to build something for Barbie: houses, cars, campers . . . no need for the pink plastic ones—we had an energetic, if amateur, craftsman in the house! Fortunately, I was well equipped with a toolbox and a basement full of wood scraps left over from my father's projects. He was an engineer, as was his father. I spent of lot of time working on his sailboats with him, and he was always happy to show me new woodworking techniques and offer suggestions. While I never did figure out how to play Barbies, my role as builder suited everyone. The Barbies were well equipped, and I appeared to be following in my father's footsteps.

What I truly wanted to be when I grew up was a detective. Investigating and solving crimes seemed the height of intrigue, and I read the entire series of Hardy Boys books, twice. Unfortunately, I don't hear well. My sisters and I were all born with congenital hearing loss. Even at eleven, I knew that I would never pass the physical exam required of police officers, much less be able to listen in on people or avoid being sneaked up on from behind. I decided instead to become some kind of scientist, another occupation that involved investigation. I spent a lot of time outside, watching the salamanders in our creek and catching tadpoles to observe them eventually grow into frogs. But it was my sixth-grade earth science teacher who inspired me to be an earth scientist.

Fast-forward ten years. I began college majoring in geology, which I loved. I was never happier than when doing fieldwork or studying crystallography. But the job market for bachelor's-degree geologists seemed slim. The 1980s was a time of prosperity, young upwardly mobile professionals (yuppies), and the pursuit of personal wealth. With parental encouragement and support, I transferred to engineering school where I graduated in 1985. I enjoyed the logic and the exacting nature of engineering, and I loved working in the machine shop. But secretly I regretted leaving geology, because I wanted to study the earth, not build things to make it more habitable. However, the definition of success as an adult seemed clear—a photo-perfect family, a high-paying job, and the acquisition of nice things. I got married and began working my way up the management ladder at a company outside Boston. By the time I had my first son, I was working sixty-hour weeks and consumed with "getting ahead." I even started attending an MBA program at night funded by my employer. I did well in classes but felt completely alienated by the environment at work and in business school. This wasn't what I wanted! I made a good salary but absolutely hated my job. The work was not technically challenging, the results seemed irrelevant, and I disliked the time- and energy-consuming office politics. I went home at the end of every day exhausted and feeling that I had nothing to show for my work. How, I wondered, did I get into this mess?

Fortunately, fate stepped in. Both my marriage and the company I worked for fell on hard times, and I was forced to make some major changes. I knew by then that business was the absolute wrong place for me. I needed to go back to school so that I could get the advanced degree that would allow me to do independent research. Now, graduate school, with its long hours and low pay, is not designed for parents, much less single ones. Had I known what I was getting into, I might have been daunted by the enormity of it. My older son lived with his father for most of the year. But my younger son was just three when I went back to school. The plus side, for him, was living in graduate student housing. For a small child, living in close quarters with other young families provided an endless source of playmates. Even better, many of them were foreign, so my son had many opportunities to befriend people from other cultures. His first best friend while I was at the University of New Hampshire was a Chinese girl. Later, after I transferred to Dartmouth College for my PhD research, his best friend was a Japanese boy. Because all the science and engineering graduate students had similar goals and challenges, graduate student housing was a supportive environment for both of us. I was usually able to work out child care exchanges that gave me the flexibility to carry out experiments at odd hours or to attend early or late meetings.

When all else failed, I took my son to school with me. Over the years, both boys spent many hours in my office drawing and building "robots" with spare electronics parts. They watched me work and asked lots of questions. Sometimes they got to talk to other graduate students or visit their labs or even hang out on campus with undergraduates from our group. They sat through meetings and occasionally classes. It wasn't a completely normal childhood, but it was a time rich with interesting experiences and opportunities for learning. Most important, perhaps, it demonstrated to my sons my commitment to learning and gave them a model for pursuing what you are interested in, even when it requires sacrifices. They have seen me define and defend my choices and my priorities. They have seen me struggle, triumph, and find my way back to what I am enthusiastic about. I am now a glaciologist, materials scientist, and engineer. My sons proudly tell people, "My mom studies ice." Even better, my older son says I am "cool."

I am the scientist in the family. I am the one who can tell you how things work, who can fix the car, and who can help you with your math homework. I not only appreciate but enthusiastically take part in schemes involving action figures and parachutes. I think Legos are the greatest invention since the wheel. I'm not the world's best cook, but I can explain the chemistry behind cooking and the difference between microwave and convection ovens. I'm clever at producing materials for odd projects and am always willing to help build something, whether it is as big as a tree house or as small as a snowboard for a teddy bear.

During graduate school I had the opportunity to mentor younger students both formally, through mentoring programs, and informally. I always told them, "If you do what you love, you will be good at it and the money will take care of itself." I tell my sons the same thing. They show no signs of wanting to be scientists or engineers themselves, but whatever they choose to pursue, I hope I have shown them how to be committed to it. My route has not been the most direct, nor the easiest. It is hard to change careers and to pursue graduate school when you already have children. My mother has supported me every step of the way, and this is a big part of my success as a scientist. I hope I can help my children achieve their goals, whatever they are.

Parsimony Is What We Are Taught, Not What We Live

Sofia Katerina Refetoff Zahed

PhD Candidate, Zoology, University of Wisconsin-Madison

We are taught in the sciences that the most parsimonious answer is likely to be the right one. However, this is not the case if you are living a dual life of mother and scientist. The simplest explanation is certainly not the most likely to be true. If it were, you would not be raising children while simultaneously pursuing a doctoral degree. The two are mutually inhibitory. Student life means working long hours and receiving little money for your efforts. Child rearing means working long hours and receiving no money for your efforts. Therefore, if you are busy in the lab and have little money, how can you pay the high costs of child care? How can you be with your needy and irresistible infant and yet work overtime on campus? The most parsimonious answer is that you cannot raise your own infants and be a graduate student in science.

Nonetheless, we do it anyway. How? How can we be there to nurse our newborn every two hours and yet continue the time-sensitive experiments simultaneously? How are we to be present to witness our little ones taking their very first steps and yet demonstrate our commitment and thoroughness to precise data collection, show up to weekly lab meetings, submit grant proposals and pursue scholarships, grade exams, attend brown bags and weekly departmental colloquiums, be present at talks given by prominent visiting professors, remain abreast of the field with current publications, present at annual conferences (out of town), prepare manuscripts, and once again take more data at all hours of the day and week? Oh jeez, it's my weekend to take observations on the monkeys again. Can someone else stay up all night and nurse my child while I'm away?

I must seem cynical, but truly I am awestruck. How have women in academia done this before me? I look to women in science around me. Many are struggling. Thank goodness I have been so fortunate as to have a supportive advisor and mentor. These are two people who care deeply about parental care (which we study in primates) and integrity and are willing to accommodate odd working hours. Academia, however, still retains the same demands regardless of our familial obligations. And it is this institution of constant productivity that does not allow for time off dedicated to family. Is it not the American way to stay open twenty-four hours a day, seven days a week? Even the allowance of six weeks' maternity leave without compensation (which is the policy for assistant professors) is deficient. Oh yes, leave that infant who cannot even lift his head and is exclusively dependent on you for food in someone else's care, somewhere. And do this with no financial compensation. Where should the money come from to pay for all the new expenses of diapers, crib, car seats, stroller, bottles, blankets, excersaucers, squishy toys, and so forth? I am not at this point suggesting that your advisor should provide for your infant's needs, but I do wish to bring attention to the lack of support for children and their needs in the United States.

Growing up, it seemed that the only worthwhile endeavor was studying and getting good grades. When I became pregnant, I assumed I would just get child care and get back to my studies. No one mentioned that our society would work against you and that professional life did not embrace family, particularly not motherhood. No one explained to me that to love your child is to love yourself is to love your work but you won't have time for all three demands. My generation of women somehow got duped into believing that we could achieve any level of career success in any field and still be Mom. Reality check: mothering is *full* time. Thank goodness for the rising number of stay-at-home dads, extended family members chipping in to help out, and parents who continue to financially support their grown-up children and, by extension, their grandchildren. I am indebted to the women before me who took on the arduous task of fighting for women's rights in the workplace; but somewhere along the way we became expected to each be "super woman." We can't do it all ourselves both at work and at home, all the time. Humans need to grow up and be part of a cooperative system in which the tasks of raising children and making a living are shared.

The requirements of a doctoral degree can sometimes be intimidating as well as demanding. I gave my first presentation at an international conference while six months pregnant. My two-year-old daughter managed to choke on a one-cent euro coin right before I delivered my talk. What timing. And the night before I had my qualifying exams, she had a fever of 104 degrees, her first ever. Behind the scenes there are constant mini-emergencies, such as falling from the playground monkey bars. In addition there are constant responsibilities, such as accompanying your child on her first bus ride or field trip and routine pediatric checkups followed by almost continuous visits for ear infections and other winter maladies. Basically, active children have daily demands that you must somehow integrate into your studies. I used the commute from campus to day care as a time to make my phone calls while breast pumping so that I had milk to provide at pickup (thus far without an accident). The pursuit of a higher degree and mothering both require time. Any way that you can combine your efforts and multitask is time well spent.

Clearly what my husband and I lack is time for ourselves and time for each other. We barely have time to talk, and we never date anymore. We don't have the time or the money to do so. We lack sleep, and I am sustained by large quantities of coffee. Our marriage has suffered. One day soon we hope to have time again for each other—when the children are older and when he has completed medical residency. Sacrifices must be made in raising children when both parents are studying and training. We at least have similar goals and love each other to the core.

Now that I have painted our desperate picture, let me tell you that our daughter is awesome. Despite the naïve choices of her parents, she is lovely, healthy, funny, kind, and brilliant. So is her younger brother. Despite our difficulties, we surround them with caring friends, family, and neighbors. My daughter and I spend weeks cooking together simply to prepare for a "because" party, to celebrate friendship, and to have an excuse to crack a piñata. We take time to fish, chase fireflies, make daily art projects, read the classics, plant a garden, and use the fruits and vegetables to make food and gifts. I do all of our grocery shopping with the children in tow, and we include them in all chores at home. In addition, the kids are included in our professional work. My daughter could lead a field trip to our primate facility, as she knows what to feed the monkeys, which individual monkey expresses the most paternal behavior, and which of the fathers are duds. And of course, she could discriminate between a monkey and an ape before she turned two. My husband and I do not go out to movies and dinner or socialize except for university-related activities, and then the children join us. In terms of work, we make up for it at night when the children are asleep. My husband and I trade off time on the weekends to allow the other spouse to work. We wake in the middle of the night to write and to read. We work on the plane or on the bus. We maximize. We prioritize. We adjust and we refocus. Clearly, we are not the first parents to juggle numerous demands. In fact, my grandparents had the challenge of raising their children in Europe during World War II. They read by candlelight, while wrapped in blankets, as electricity and fuel were scarce. They scraped by with little means. They maximized their time. They prioritized. I do not mean to compare my challenges with those of people who lived during wartime. I mean only to say that we are human. We optimize. We adjust and assimilate. We are ubiquitous for a reason. We reproduce and produce.

What is the most parsimonious answer? Don't have babies while you are in a graduate program! That's actually pretty obvious, and most people figure it out. Academia does not get less demanding, however, and thus post-doctoral fellows and tenure-tracked assistant professors are in the same boat. The ticket is to have supportive extended family nearby. Involved grandparents are lifesavers. Unfortunately, most of us cannot take advantage of this because we move to wherever the job or the facility that allows our specific type of research is located. In that case, beg, cajole, and persistently bribe your mother to move nearby. Make sure she knows that she is welcome to stay for extended visits.

My work centers on understanding the triggers that lead to invested parental behavior in two species of New World primates and by extension those for human parents. All family members of cotton-top tamarin and common marmoset monkeys engage in extensive parenting behavior, including carrying, protecting, and feeding often twin infants. Like the tamarins and marmosets, we raise our young for selfish reasons: to perpetuate our genes, to promote our species, and to enrich our lives. Although some male tamarins and marmosets spend loads of time with their young and greatly contribute to raising their offspring, some don't. Some human fathers are highly invested in their young while some are not. Wouldn't it be amazing to figure out the reason why? Wouldn't it be helpful to know beforehand which males were going to stick around and help teach and nurture their children and who would leave before the labor pains began? In a world where human populations continue to grow and children are left without caregivers as a result of war or illnesses such as AIDS, understanding the drive to care for both related and unrelated young is crucial. Although this is a simplified explanation of cooperative breeding systems of humans and monkeys, it gets at the main point. It's rewarding to make new discoveries, and results from research can have practical and beneficial applications.

If pursuing graduate studies while becoming a mother is such a struggle, why do it? The truth is that I love research and academia. Being constantly challenged to dream up new experiments and new ways to ask questions and just getting one step closer to an answer are invigorating. In addition, the most valuable lesson I can give my children is to provide a role model. Between college and returning for a PhD I taught in numerous venues such as science museums, public schools, and even on a public access television show called *Science Power.* Yet I couldn't wait to get back to the university in order to do research and to be in an environment where people all around you are driven to obtain new knowledge, to share information, and in a basic sense to make the world a better place. Perhaps this is idealistic but colleagues near and far repeatedly affirm this view. It's a remarkable community to be a part of, one that I would like my children to grow up in. There is a way to contribute to academia and to raise children. It will require a major shift in view of the *pace* needed to fulfill requirements for a successful academic career. Can you imagine slowing down the clock for one, two, or even five years, allowing infants to be nurtured and to grow, and then resuming your studies with added experience, savvy, insight, and motivation? Wouldn't it be helpful if we could organize a temporary leave of absence and return to our studies without being penalized by, for example, having to re-apply to the program and paying re-entry fees? As noted in the Stanford Graduate Student Handbook, it is important to acknowledge that a woman's prime childbearing years are likely to be the same years that she will be studying in graduate school, pursuing postdoctoral training, and establishing herself in a career. Whether the pause in studies means taking one to five years out for kids and then picking up again right where you left off at graduate school, or finishing graduate work and then taking a few years to start raising a family before pursuing a postdoctoral position, it means slowing down and taking time off for your children. Although there is much that can be done toward your degree even while caring for an infant, such as analyzing data, writing manuscripts, or attending talks, it still means slowing down the process.

Let's move from theory to practical ways to implement a new timetable. Professors (academic advisors) already pay a premium for graduate research time by covering their tuition and stipend, costing approximately $35,000 per student annually. In addition to graduate tuition and salary, advisors must cover hourly wages for undergraduates, support staff such as animal care, and the direct animal husbandry costs, all on soft money. With the nationwide squeeze in grants funding, the ability to meet these expenses is shrinking. Financial assistance to allow graduate students to have families—or in other words, to avoid limiting the graduate body to childless students—should come from the universities, who ought to receive funding from both the state and the federal government. This is a radical challenge; however, we can begin simply by changing our view about mothers in science. Women are a valuable part of society, and, given the extra time to be both mothers and scientists, we will add value to campus life. If I can achieve being a mother and a scientist, perhaps my daughter can imagine having a career and a family. Perhaps in another generation we can make the necessary changes to make that a reasonable goal.

In Denmark, where I spent many of my childhood summers, some mothers can take fifty-two weeks of maternity leave and are often given a full salary while at home with their infants. Women are commonly given four weeks leave before the due date as well. Subsistence allowance for parents and children is awarded by the municipality of residence but the specific pay and number of weeks for maternity leave postpartum are dependent on the individual's past employment. Fathers, too, have the option for extended paternity leave. It is clear that in Denmark, children are important. Pregnant women can park near the entrance to a place of business, and public bathrooms may be equipped with a holster for toddlers on the stall door so that parents can safely have their hands free to fasten their pants (also available in Scandinavian IKEA stores in Schaumburg, Illinois), and people willingly give up their place in line for pregnant women or parents with children. How often do you see that happen in the United States? What is it that we value in the United States? Our national policy is to allow six weeks' maternity leave with no pay. But, the difference I would particularly like to highlight is that in the United States, women who take the time for their children are frowned upon rather than celebrated. What our political and educational systems value above all else is production of a product or, in the case of academics, of publications. It's the bottom line. What is not considered is how the bottom line is attained and what sacrifices are made to accomplish it.

Although change on a national level may take time, we can advocate right now for change on our campuses to reward and support women with young children. We (three graduate students in microbiology, zoology, and medicine) are currently working to establish a policy for family leave for graduate students on our campus (whether for maternity or paternity, to assist in handling adoption or foster care placements or to care for a sick parent or spouse). Our goals are similar to what has been developed in a few progressive departments, such as chemistry at University of Wisconsin at Madison, and on a few campuses, such as Stanford University. Observations have shown that tamarin and marmoset mothers are intensely invested right after parturition but that assistance from other family members, once infants have matured, diminishes maternal contribution and benefits everyone involved. Similarly with humans, given a few years to mother our infants, we can return to our studies with vigor. Mothers in science are amazingly efficient. We have a wide breadth of experience. We are dedicated. We are motivated. We care about our jobs. We care about our families. We want to contribute to society, and we have the skills to do so. Don't shut us out. Embrace us. We will complement our fellow scientists.

Role Models

Out with the Old and In with the New

Marie Remiker

Fourth-Year PhD Candidate, Zoology

After two years of graduate school, I fell in love with the man of my dreams. He was an ecologist, and my research field station was located in the city where he lived and worked. We met through colleagues, and two years later we were married. We had a cozy three-bedroom home in a small town, but I didn't live there. I lived in a tiny one-bedroom apartment about two hundred miles south of our home, where my university was located. After two years of commuting (seven hours of driving every weekend) during the school year, we decided we had had enough. We wanted to live together like a normal couple. My husband's employer was very accommodating, allowing my husband to move to be with me while I continued my education. My husband's employer even paid for office space for him to work! By the end of the semester I had finished all my required coursework, so logically we decided to move back north. Because of our move, there would be no extra financial costs for employers, and we could save money by getting rid of my apartment. My husband's employer was willing to give me space to work on my dissertation, and my field site was only minutes away from our house. I thought our plan was wonderful. My husband and I would be happy because we could finally be together on a daily basis, and I could easily continue my research from the new location. Then I spoke with my advisor.

He did not want me to move. However, because I had my own funding, he told me the decision was ultimately mine. He added that he would not have hired me if he had known I would make this move and informed me that if he were supporting my stipend, he would not pay me if I moved away.

I was blown away! How could he not understand? He was a family man, or at least that was what the other graduate students (all men) had said when I was deciding whether to join this lab. He must have known how important family was to me. A few months earlier, there had been a death in my family that hit me pretty hard. My advisor seemed sympathetic during that difficult time. Why did he not understand now? In order for my relationship with my husband to grow as it should, we needed to be together. My advisor was married, so I assumed he would understand, but he did not.

Eventually he told me he would support whatever decision I made, but his earlier statement completely contradicted that. Which one do you think I was going to believe? How could he regret hiring me and yet support me in moving away? And what did "support" mean anyway? He did not currently support me financially, so he must have meant he would maintain the advisor-student relationship we had. After our conversation, I was hurt by his comments but felt lucky that he was willing to support me in my decision. Looking back on it, I am angry that I felt lucky. I feel that my advisor should have accepted my decision without making me feel guilty for having made such a difficult choice. I was simply trying to balance my attention to both my family and my career. To do so, I would be moving about two hundred miles north of my university's campus and working in a lab with equipment and facilities similar to those I had available in my advisor's lab. I had many of the resources necessary to continue my research at this new location. My husband's employer was willing to give me space and access to these resources for no charge. Why, then, did my advisor resist my move? He said I wouldn't be getting the full graduate student experience living so far away. I wouldn't be immersed in the academic atmosphere that I needed to fulfill my graduate career. I found it hard to believe that he was serious. By that time, I had been immersed in the "academic atmosphere" for more than six years! I was happy to think I wouldn't have to battle forty other students for the last bit of standing space on the campus bus in the middle of a blizzard.

Maybe his definition of "academic atmosphere" was a little different from mine. Even so, I think there was something more to the whole disagreement, a deeper reason for his objection to my move that perhaps he didn't even recognize. Maybe he was worried that I would drop out of graduate school altogether and therefore prove to be a waste of time and resources. Perhaps he worried that I would disappoint him as his previous female graduate students had. I have to admit that I was a little worried too. When I interviewed at his lab, my advisor told me that he had had "bad experiences with women from small, liberal arts institutions." He said that these women (I like to call them SLI women) did not seem to adjust well to the large university life. I now believe that he could just as well have said, "I have had bad experiences with graduate student women," SLI or not.

Currently I am the only female in our lab—seven men and one female, and my advisor regrets hiring me. I sometimes wonder whether he regrets hiring all women. I realize he regrets hiring me because I put my family first, and I am deeply saddened by this realization. At the moment, my husband is my family. How many women have gone through a similar experience? How many women must face the fact that they have disappointed their mentor and role model because they made a choice to put family first? I have hope that some day the balance between family and career will not be so difficult to juggle and that choosing to put family first will be a decision that is respected and not scorned. Maybe one day I will have the courage to tell my advisor how much *he* disappointed *me* on that day—the day he made me feel guilty for choosing my family over my career.

As a woman and scientist-in-training, I often worry about properly balancing my life. My husband and I are considering having children. My sister, who is an obstetrician and gynecologist, tells me that childbearing complications increase rapidly after the age of thirty. Can you believe it? I am twenty-six, and I hope to have at least two children! "Well, honey," I say to my husband, "we'd better get a move on this." After all, I don't need age to complicate factors. Having two scientists for parents will make our children strange enough! In all honesty, I fear having children because I am unsure of how I will provide the necessary time and attention for my children while maintaining a successful career as a scientist. Will my employers understand and grant me maternity leave? My earlier experience with my advisor only reinforces my fears. When considering children, I worry about disappointing him yet again. Despite these concerns, I know there are successful women who have managed to balance their families with their careers in science, although these women are still somewhat rare.

I have a friend who is a single mother and is working on her PhD degree in zoology while pursuing a master's degree in social work. She is one of those rare women, and I wonder how she does it. She tells me that her interest in social work is to help young, single mothers like herself because being a single mom *is* tough. Still, I wonder how anyone, let alone a single mother, can raise children while attending graduate school in the sciences and not feel overwhelmed. She is a wonderful role model because she has overcome many obstacles despite the difficulties of being a woman and mother in science. I often worry whether I can manage these two roles as successfully as she has done.

Unfortunately, my school has no policy on maternity leave for graduate students. I contacted our Women in Science program director about this, believing that the university must surely be working on implementing such a policy. She told me that graduate students typically work this out with their advisors, most of whom are very understanding. I think about my advisor, however, and am not comforted. After my experiences as a graduate student thus far, I have realized the importance of a good mentor and role model and have decided my next mentor will be female. I feel that my current advisor cannot relate to my concerns, my feelings, or my obstacles as a woman in science. I don't know for certain, but I suspect he would encourage me to wait to have children until I finish graduate school. Somehow I do not believe life as a postdoc or junior faculty member will be any better suited to having children.

When is the best time to have children when you are pursuing a career in academia? I've heard many different opinions. Some think graduate school is best because the schedule is so flexible—these people must have forgotten the financial woes of being a graduate student, or maybe they lived in a warm climate where being homeless is not so bad. Others feel that after tenure is best because you then have stability in work, pay, and life in general. These people don't know that I am moving at snail's speed, and by the time I have tenure, I won't be thinking of having children but how best to avoid hot flashes! Honestly, I will probably be only about forty by that time, and my sister will probably encourage me to adopt. Though no one I have talked with has recommended having children as a postdoc, maybe I will give this one a try so I can comment with wisdom someday when some unenlightened female graduate student asks me for advice on the subject.

I'm guessing there really is no "best" time to have children when pursuing a scientific career in academia. Life as a female scientist is difficult, and the challenges of being a mother and a scientist can be brutal. I can only hope that when I become a mother, I will have the strength and ability to pursue the career I want as well. I don't feel I should have to give up one or the other. I feel it is my right to do both, to be a mother and a scientist. Perhaps one day advisors, male and female alike, will agree with and support this right as well. I dream of a job in academia where I can teach, do research, have access to on-site day care where I can visit my child to nurse or check on him or her as I please, and be home during the evening hours and on weekends to be the mother and wife I would like to be. Is this possible? Realistic? Is there a balance between work and family that can be maintained by a female scientist in academia? Will I have to choose to opt out? Will this disappoint my spouse, my mentors, or worst of all, me? How does one balance career and family? How does one not feel guilty for the decisions she makes regarding children and providing child care?

The answers to these questions can be found in those women who are making it in the world of science and the world of motherhood today. Although I have yet to learn many of their names, these women are my new role models. They will become my mentors as well. They must, in order for me to succeed in my career choice. My advisor is not the role model for me because he cannot relate and perhaps never will. Maybe one day he will learn from me about the challenges of being a woman in science so he can properly mentor a female graduate student. This knowledge is just as important for an aspiring female scientist as the scientific knowledge he has to impart. If we want to increase the number of women in scientific academia, we need to understand and support the many roles a female scientist may play in today's society, including both wife and mother.

Pursuing Science and Motherhood

Kimberly D'Anna

Fourth-Year PhD Candidate, Zoology, University of Wisconsin-Madison

I am a researcher, a student, and a mother. I am like any other graduate student in that I study for courses, attend seminars, and investigate research questions. However, I am unlike most students in that I collect food stamps and child care funds from the county in order to support my son and myself. I spend much of the day planning every minute after 5:00 p.m., from arriving at day care on time to dinner to arranging activities to attempting to enforce bedtime. Each morning I try to arrive in the laboratory by eight o'clock to begin a day's worth of work efficiently, but often I don't arrive until nine because I can find only one of my son's shoes, he has let the cats out of the house, or I forgot to make his lunch. In the laboratory I perform precise cannulae surgeries and run assays, but at home I find it hard muster the energy to play a game of T-ball or read bedtime stories. I am a twenty-five-year-old single mother of a four-and-a-half-year-old and a graduate student in the field of behavioral neuroendocrinology.

It seems odd to mention these two professions in one sentence, as I have, probably because they sound opposing, but they are the makeup of my life. I am lucky that my advisor is supportive of my limited schedule. He is part of a dual-academic-career couple, and one of his two boys is a month older than mine. When I was looking for an advisor, my priority was finding someone who would understand my situation or had a family; this was even more important than the type of research he or she was doing. Other students may find it shocking that I would choose a lab for reasons other than research, but my priorities are different, and this has sometimes posed a problem.

Though my advisor has been fantastic and I love the research that I do, the academic culture, both institutional and social, has been cold to us. During my third year I was awarded a fellowship, which was a blessing, but in order to receive it I was required to attend a summer program. I was excited but soon found out that the facility would not house students with children. I didn't have anyone who could keep my son during the month-long course, and I could not bring him. In the end, I was able to retain my fellowship without attending the program, but I missed out on a career-developing opportunity. I encounter the same problem with conferences that don't provide child care. I am fortunate enough to have parents willing to watch my son, but women with families who can't attend these conferences, which are so vital to networking and the sharing of research, are at a disadvantage. I have also run into complications with others in my department, especially the other graduate students. I have not become acquainted with any of the students in the department other than my lab mates, in part because they are never willing to gather at a family-friendly place. If I want to do anything with the department, I must bring my son, but when he is around, other graduate students ignore, avoid, or exclude us from activities. I have stopped attending most department functions, and my only friends tend to be other single parents.

I have found that support is the key to my success in science. When studying for my qualifying exams, I had little support and was expected to carry on with research, coursework, and my parenting duties as normal. I cried almost every night for a month because I could not balance the load. I would do research, attend classes, and study what I could during the day, and then I would attempt to study more when I got home. But naturally my son wanted to be with me. I would yell at him out of frustration and desperation. Sometimes he quietly retreated to his room, but more often he cried, yelled, pulled on me, and begged for attention. I would feel awful. I have never felt like a worse person or mother. This went on for about a month, and when I got to my qualifying exam, I broke down in tears after an hour in front of my committee. I was horrified that I had put an exam above my child's well-being. I began to wonder if staying in science was selfish and if I could balance it all or was even cut out for it. But then I realized that everything was just going to take me longer than the average graduate student because I don't fit the typical profile—I never have. I had my son as an undergraduate when I was twenty years old and have been struggling ever since.

For a while I resented everyone without children and even those with children who were married and settled. From my perspective in the academic sciences, "settled" is a term that applies to the far-off future. After I finish my graduate work, I will be off to a postdoc, then possibly another, and then to who knows where for an academic position. I will be uprooting our lives for the next five to ten years. But then again, there are really wonderful things about being a woman in science. As a scientist, I have a flexible schedule; as a mother, I am forced to have a life outside the lab, which gives me perspective. I also know that furthering my education can only help my family in the future. I do it because I love my son and I enjoy the work.

I don't know how long I will stay in science. I can only hope that as more women with children enter the academic sciences, science will become more accommodating to our needs as mothers, which ultimately will increase the quality of the work. I also hope to give the message to women in science who have children or are thinking about having children that it can be done, and done well. This was something I was unaware of when I entered the sciences. I assumed most women in science chose not to have children and that female scientists with families were rare. I now know differently and am so happy to meet other mothers in science because we need to know that even if we feel we are alone, we're not.

Conclusion

Among the tens of thousands of women in the United States who have turned their passion for the sciences into a profession, the thirty-four contributors to this volume are professors, research scientists, grade school teachers, and writers; they mentor graduate students, turn children on to science in the local park, translate complex scientific research for politicians and the lay public, and inform policy; many also juggle day care (or did at one time) and quality time with their children and loved ones.

The women who penned these chapters were once girls who mucked around in swamps, wondered about invisible worlds, or pondered the night sky. Theirs were inquisitive young minds destined to pursue science, and as they did, science became more than a job—for many it is a lifelong quest. But lives quickly became complicated as careers were combined with spouses or partners and then with growing families. Although for some this means leaving scientific aspirations by the wayside, it did not for those whose stories are contained here. Whether following traditional paths laid out by academics and researchers before them or venturing off and finding their own way, all these women—myself included—have continued to nurture and grow their scientific selves.

This morning, with the kids off to school and the dog walked, as I washed the breakfast dishes before heading up to my home office, National Public Radio replayed part of Gloria Steinem's address at Smith College's 2007 commencement. I turned the water off and stood by the radio so I could hear what she had to say:

> My generation often accepted the idea that the private/public roles of women and men were "natural." Your generation has made giant strides into public life, but often still says: How can I combine career and family?
>
> I say to you from the bottom of my heart that when you ask that question you are setting your sights way too low. First of all, there can be no answer until men are asking the same question. Second, every other modern democracy in the world is way, way ahead of this country in providing a national system of child care, and job patterns adapted to the needs of parents, both men and women.
>
> So don't get guilty. Get mad. Get active. If this is a problem that affects millions of unique women, then the only answer is to organize.[1]

I realized that in Steinem's words was an echo of the stories in this book. Each succeeding generation of women has benefited from the one that came before. Women like Joan, Marla, Marilyn, and others—who, as scientists, women, and mothers, were trailblazers—started their careers at the beginning of the push toward equity in the workplace. Some became deans, while others, like Marilyn, were turned away or discouraged by academia yet kept their careers alive by finding soft money, working on a project-by-project basis, or working without pay, thus continuing to study and publish. Those of us in the next generation who followed women like Marla and Marilyn, graduating with PhDs in the late 1970s and beyond, belong to the cohort of women defined by Claudia Goldin as the "Career and Family" cohort.[2] Many of us were encouraged to do it all, and some did, for better or worse, some blissfully unaware of Aviva's discovery that it takes courage to "face the reality that you cannot be a supermom and a superscientist at the same time because, temporarily, something's got to give," and others, acutely aware of the trade-offs.

If the essays contained in sections III and IV of this book, written by successors of the "Career and Family" generation, are any indication, we

1. Gloria Steinem, Smith College 2007 commencement address, http://www.smith.edu/commencement/.

2. Claudia Goldin, "The Long Road to the Fast Track: Career and Family," ANNALS AAPSS 596 (2004): 20–35, 26.

have come a long way toward realizing flexibility in both the workplace and in the home. Some chose full-time work, some pursued part-time work, while others carved out flexible positions to suit their needs. We have come a long way, but we still have a way to go. Most married women in science are married to other scientists and engineers;[3] those who are not often have partners or spouses working a full-time job in another field.[4] Therefore, many families face not only the challenge of finding two suitable jobs in the same location but also the necessity of determining which job takes precedence when time off is required for family or child care. For some, combining family and career remains a puzzle requiring creative solutions.

Yet the continuity of service and experience—and of published work, especially in the sciences—required to build a career, any career, means we must better address the reality of women as childbearers, and often as primary childrearers. While some are content to return to full-time work soon after becoming a mother, others are not, and their devotion to a child does not mean that they are ready to leave science behind. Of those who women do choose to opt out and then opt in or who choose to pursue part-time work or flexible work options, some are bold enough or confident enough to ask for an extended maternity leave, or to work from home one day a week, or to cut back on their hours. But for some, these options are not a feasible or even desirable course of action. Though several contributors write of wonderful child care—including the support of extended family, au pairs, and on-site day care—enabling them to work the hours best suited to their career choices, just as many bemoan long waiting lists, financial constraints, and unsatisfactory situations. When parents need to put their newborn on the waiting list for a university day care slot that might be two years away, there is a problem. Some women from preceding generations with grown children write that they wish they had had more time with their children when they were younger. Today, many more are choosing that time, taking their chances that the scientific community will modernize and welcome untraditional career paths.

It is clear from the many different ways that these thirty-four writers portray their contributions to science that, though their methods of doing science might change, their dedication does not, resulting in a host of new avenues pointing to careers outside the traditional arena of academia.

3. National Academy of Sciences, National Academy of Engineering, Institute of Medicine of the National Academies, *Beyond Bias and Barriers: Fulfilling the Potential of Women in Academic Sciences and Engineering* (Washington, DC: National Academies Press, 2006), fig. 5-3.

4. Ibid., fig. 5-2.

Steinem spoke of "flexible job options." This may be one of the most important and immediately available changes that our society can make to attract and retain family-oriented workers. Though many sectors already provide for flexible schedules, including part-time work, flextime, and telecommuting, a scientific community that does not judge commitment by the hours worked but rather by quality has not yet been fully established. Those who ask for time off or reduced hours are concerned (sometimes rightly so) about being mommy-tracked or considered less serious about their work.

Traditional as it is in other respects, because of the paucity of women in faculty positions, academia is recognizing not only the need for change in work policy but that those who choose to pursue positions other than full-time tenure-track (such as part-time tenure-track) must be respected and treated as integral members of the faculty.[5] A recent report from the Center for the Education of Women at the University of Michigan entitled *Designing and Implementing Family-Friendly Policies in Higher Education* concludes:

> At some point in their lives most men and women faculty—faced with exigencies ranging from the joy of new parenthood to the challenges of ill, injured or dying family members—will desire more flexibility in their academic careers. Employers who recognize the commonality of these life events and enact effective policies to temporarily modify duties and lessen pressures are making their institutions more humane, more productive and more attractive to potential faculty.[6]

The American Association of University Professors (AAUP) devotes a section of its website to "Balancing Family and Academic Work," noting that it has recognized the need for flexible work policies for decades.[7] And since

5. American Association of University Professors, "Balancing Family and Academic Work," http://www.aaup.org/AAUP/issuesed/WF (accessed May 30, 2007); "Part-time Tenure Track Policies: Assessing Utilization," http://www.engr.washington.edu/advance/policies/WEPAN-2004-Part-Time-Tenure-Trac(accessed May 30, 2007).

6. Gilia Smith and Jean Waltman, *Designing and Implementing Family-Friendly Policies in Higher Education,* Center for the Education of Women, University of Michigan, Ann Arbor, 2006, http://www.umich.edu/~cew/PDFs/designing06.pdf.

7. American Association of University Professors, "Statement of Principles on Family Responsibilities and Academic Work," http://www.aaup.org/AAUP/pubsres/policydocs/contents/work fam-stmt.htm. For "Resources on Balancing Family and Academic Work," see http://www.aaup.org/AAUP/issuesed/WF/resources.htm?wbc_purpose=Basic&WBCMODE=Presentation Unpublished.

2001 the National Science Foundation has awarded approximately $100 million dollars to researchers and universities around the country through ADVANCE: Increasing the Participation and Advancement of Women in Academic Science and Engineering Careers. It is estimated that of those ADVANCE awards devoted to institutional transformation (which accounts for the largest proportion of the total budget), development of family-friendly policies is a central issue for perhaps as high as 90 percent of the grants.[8]

In addition to flexible work hours is the issue of those who choose to step out of academia or away from the bench for several years while their children are young, a situation that in the past often resulted in a permanent exit from research science or science altogether. Recognizing that those who step away represent a potentially valuable population, the National Institutes for Health (NIH) encourages both men and women who have taken time off to care for children or other family members to resume their research careers, stating in its program announcement:

> Among the reasons for the low representation of women may be the fact that women bear a majority of the responsibilities surrounding child and family care. To address this issue, this program is designed to offer opportunities to women and men who have interrupted their research careers to care for children or parents or to attend to other family responsibilities. A second objective of the program is to mentor and guide those who receive support to reestablish careers. . . .[9]

Finally, although there are not yet formal programs or movements that I know of that support and encourage those who choose to develop science careers outside academia and away from the bench or the field, it would be interesting to fully identify the numbers of women (and men) who would place themselves in this category, their reasons for this choice (data from the National Academy of Science suggest that for many women, family considerations are important), their degree of satisfaction, and if possible some measure of their combined contribution to science as a whole.

8. The ADVANCE portal website is loaded with reports, abstracts, speeches, tutorials, and other information from colleges and universities from across the country on everything from policies that influence change to dual careers to family care. ADVANCE portal website, http://research.cs.vt.edu/advance/index.htm.

9. "Supplements to Promote Reentry into Biomedical and Behavioral Research Careers," http://grants.nih.gov/grants/guide/pa-files/PA-04-126.html.

If the scientific establishment is serious about retaining young parents—and I do believe it is—then universities, government and nongovernmental organizations, and the science industry in general must support affordable, quality child care for their scientists. They must recognize that time off for a child is not necessarily matched by diminished dedication to work; in some cases it may even spark new scientific aspirations leading to greater societal benefits. A woman's childbearing status alone should not impair her prospects for a fruitful career. It is time, as Steinem declares, for women's—and more specifically, mothers'—decisions to be informed by their own desires and not by any "natural" public or private role dictated by society.

There are tens of thousands of us women scientists, and we know that childhood is fleeting and life is short. If we want change, the time is ripe to speak out for what we want. In this age of Internet-based information, organizing is more feasible than ever before. There is already momentum toward greater job flexibility and improved child care, so let's keep it going. The possibilities examined in these thirty-four essays are just the beginning. They are the foundation for an ongoing forum in which others can share their own stories, ideas for change, advice, and reflections on the dynamic environment that is the scientific workplace in the twenty-first century. We invite the reader to participate in continuing discussions devoted not only to outing the elephant in the laboratory but also to welcoming the future generations of scientists who give us hope and purpose.

Contributors

A. Pia Abola is a biochemist and the mother of a nine-year-old son and twin five-year-old daughters. She received her BA in chemistry from Princeton University and her PhD in molecular and cell biology from the University of California, Berkeley. She is currently part of a multidisciplinary effort to develop a quantitative understanding of how cells respond to extracellular signals. She lives in the San Francisco Bay Area, an exceedingly difficult place to leave, with her husband and three children. She can be reached at apabola@sbcglobal.net.

Cal (Caroline) Baier-Anderson is a health scientist with Environmental Defense and an assistant professor in the Department of Epidemiology and Preventive Medicine at the University of Maryland, Baltimore. She is the mother of two teenage boys. At Environmental Defense Cal provides technical and scientific support on chemical regulatory policy, air toxics, and nanotechnology. Cal earned a PhD in toxicology in 1999 from the University of Maryland, Baltimore, where she also provided technical outreach assistance to communities living adjacent to hazardous waste sites. As an assistant professor, she teaches a course in risk assessment, drawing on her experiences working with communities and as an industry consultant conducting risk assessments on chemicals found in drinking water. She has also consulted for the state of Delaware on air toxics health risks.

Joan S. Baizer lives, works, and shovels snow in Buffalo, New York. She is an associate professor in physiology and biophysics at the University at Buffalo and the mother of a nineteen-year-old developmentally disabled child who lives in a nearby group home. Her research is on the functional organization and comparative anatomy of the visual and vestibular systems. In addition to teaching and research, she is involved in disabilities advocacy for children and their parents. She received her BA from Bryn Mawr College in 1968 and her MS and PhD from Brown University in 1970 and 1973, respectively. She did postdoctoral training at NIH from 1973 to 1976.

Stefi Baum and her husband and collaborator, Chris O'Dea, live with their four teenage kids, three dogs, and three cats in Rochester, New York. She is director of the Chester F. Carlson Center for Imaging Science, an interdisciplinary university research and education center at the Rochester Institute of Technology. She received her BA from Harvard, where she co-captained their soccer and lacrosse teams, and her PhD in astronomy from the University of Maryland. Following a postdoc in Holland and a year as a Hubble Fellow at Johns Hopkins University, she moved to the Space Telescope Science Institute, rising to lead the engineering division, before taking a Science Policy Fellowship at the U.S. Department of State. She has written more than 130 refereed papers. In her spare time, she gardens.

Aviva Brecher holds a PhD in applied physics and is a federal government researcher. She has worked for more than thirty years on a broad range of topics in academia, government, and business consulting, including a year as Congressional Science Fellow in the U.S. Senate. Born in Romania, she immigrated to Israel, married an American while a student at the Technion, and transferred to MIT, completing her PhD at University of California-San Diego. She has been happily married to an academic physicist for over forty years, and they have two grown children.

Teresa Capone Cook is a high school biology teacher and mother of two daughters, aged thirteen and ten. She received her BA from the University of Colorado-Boulder in 1985, her MS in ecology from East Texas State University in 1988, and her PhD in biology from the University of North

Carolina-Chapel Hill in 1994. She has taught biological sciences and designed science curriculum for many ages—preschoolers, fifth-graders, high school students, college students, and continuing education students—and in a wide variety of settings—environmental education centers, Duke University's Talent Identification Program, a magnet high school, private primary schools and high schools, and at several universities. She lives in Woodstock, Georgia, with her husband and two children. She can be contacted at American Heritage Academy, where she currently teaches high school biology, zoology, and botany part-time, at tcook@ahacademy.com.

Kimberly D'Anna is a fourth-year graduate student in the PhD zoology program at the University of Wisconsin-Madison. She received her BS at Michigan State University in zoology with a concentration in animal behavior/neurobiology and a specialization in Chicano/Latino studies. She is also pursuing a master's in social work in the hope of combining maternal behavior work in the field of behavioral neuroendocrinology with social work to more fully address the needs of those being studied. She has a four-and-a-half-year-old son named Manuel whom she had at twenty years old during her undergraduate studies. While she loves both her work and her son, she is still looking for that perfect balance; perhaps it doesn't exist, but she is willing to try to find it.

Carol B. de Wet received her education at Smith College (BA, English and geology), the University of Massachusetts (MSc, geology), and the University of Cambridge, England (PhD, geology). While at Cambridge she met and married her husband, also a geologist. She completed a postdoc at the University of Kentucky and had the first of three children. At Franklin & Marshall College she and her husband share a faculty position. Carol also has a new position as Special Assistant to the President and Provost on Women's and Family Issues to research ways to help faculty balance their professional and personal lives. She received the Biggs Award for Excellence in Teaching from the Geological Society of America, has received Petroleum Research Fund and National Science Foundation grants, and has publications in geology and on women and science.

Anne Douglass is an atmospheric chemist and mother of five children, aged twenty-two, twenty-eight, thirty, thirty-two, and thirty-five. Two of her daughters, Elizabeth and Katherine, have chosen scientific careers and also contributed essays to this book. Anne earned three degrees, all in physics: a BA from Trinity College in Washington, DC, in 1971, an MS from the University of Minnesota in 1975, and a PhD from Iowa State University in 1980. When she isn't visiting her grandchildren or on NASA travel, she lives in Silver Spring, Maryland. She can be contacted at Anne.R.Douglass@nasa.gov.

Elizabeth Douglass, Anne's second daughter, is a graduate student at Scripps Institute of Oceanography. She also attended Villanova, earning her BS (physics) in 2000. Elizabeth spent a year as an Americorps volunteer working with Habitat for Humanity and expects to complete her PhD in 2007. She has spent time at sea and studied in Germany, but for now home is in La Jolla, California. Contact Elizabeth at edougl01@aol.com.

Katherine Douglass, Anne's oldest daughter, is an emergency room physician. She received her BS (biology) from Villanova in 1998 and her MD from Georgetown University in 2002. She is presently a fellow at George Washington University, working on an MS in public health and international medicine. Katherine spent time in 2007 in South America, Africa, and the Middle East; she calls Washington, DC, home. Contact Katherine at k.douglass33@gmail.com.

Deborah Duffy is an animal behavior research associate and mother of a four-year-old daughter. She received her bachelor's degree from Lafayette College in 1996 and her PhD in psychology from Johns Hopkins University in 2001. She began her career as an animal behavior researcher studying reproductive behaviors in wild birds. Currently she studies the behavioral development and temperament of domestic dogs at the Center for the Interaction of Animals and Society in the School of Veterinary Medicine of the University of Pennsylvania. She and her husband both grew up in Bucks County, Pennsylvania, where they are now raising their daughter.

Rebecca A. Efroymson is an environmental scientist at a U.S. government research laboratory. She received her BA in biology and English from LaSalle University in Philadelphia in 1987 and her MS and PhD in environmental toxicology from Cornell University in 1990 and 1993, respectively. She currently works in the field of ecological risk assessment and works less and less on chemical toxicant issues and more on habitat loss and physical stressors to wildlife (e.g., noise and collisions with wind turbines) for sponsors such as the U.S. Department of Defense, the U.S. Department of Energy, and the Bureau of Land Management. She is married to a scientist, Bill Hargrove, has a four-year-old son, and had a daughter after her essay was written.

Suzanne Epstein is an immunologist at the Center for Biologics Evaluation and Research of the Food and Drug Administration. She does research on influenza virus vaccination as well as regulatory review, policy work, mentoring, and research management. She is the mother of two sons aged twenty-four and twenty-one. She received her BA in chemistry from Harvard University and her PhD in biology from the Massachusetts Institute of Technology in 1979. She also has extensive experience as a performing musician. She lives in Bethesda, Maryland, with her husband and can be contacted at epsteinsue@aol.com.

Kim M. Fowler is a senior research engineer at Pacific Northwest National Laboratory. She has two boys in elementary school and a wonderful husband of sixteen years. She received an MS in environmental engineering from Washington State University in 1996 and works in the fields of sustainable design and pollution prevention. She also teaches one class every other year at the local university. Ms. Fowler's research requires fieldwork, which involves ten to twenty business trips a year. She and her husband keep busy volunteering in their boys' schools and being engaged in their boys' activities, which include gymnastics, baseball, Scouts, Lego league, dance, and fun runs.

Debra Hanneman received her BS in natural history from the University of Toledo, Ohio, in 1975; an MSc in geology from the University of Calgary, Alberta, Canada, in 1977; and a PhD in geology from the University of Montana, Missoula, in 1989. She taught geosciences at Mount Royal College in Calgary, Alberta; at Montana Tech of the University of Montana in Butte, Montana; and at Montana State University, Bozeman, Montana, from the late 1970s to the mid 1980s. After teaching at the postsecondary level, she started her own consulting company, Whitehall Geogroup, Inc., and a website, Earthmaps.com. She continues work with both her company and her website today. Her major research interests are in continental sequence stratigraphy and paleosols.

Deborah Harris is an experimental particle physicist at Fermi National Accelerator Laboratory in Batavia, Illinois, and the mother of two children aged six and ten. She received her BS from the University of California,

Berkeley, in 1989 and her PhD from the University of Chicago in 1994. She has spent the years since her PhD working on various experiments on neutrinos, which are some of the most plentiful yet least understood particles in the universe. She is currently the project manager of a new experiment called MINERvA (*http://minerva.fnal.gov*), which aims to get the most complete picture yet of the way neutrinos interact in matter.

Andrea L. Kalfoglou is a social scientist and bioethicist. She received her BA in political and social thought from the University of Virginia and her PhD in health policy and bioethics from the Johns Hopkins Bloomberg School of Public Health in 1999. She has more than fifteen years' experience designing and managing qualitative and quantitative research studies and health-policy projects. She is currently an assistant professor at the University of Maryland, Baltimore County. She is an expert in reproductive and genetic ethics, women's health issues, and human research subject protections. She lives in Potomac, Maryland, with her husband and two sons (six and three years old). She can be contacted at akalfogl@yahoo.com.

Marla S. McIntosh is a professor of plant sciences at the University of Maryland, where she is a Distinguished Scholar-Teacher and a Fellow of the Academy for Excellence in Teaching and Learning. She also has the honor of being a Fellow of the American Association for Advancement of Science, the Crop Science Society of America, and the American Society of Agronomy. She married Kevin McIntosh in 1979 and has a twenty-three-year-old daughter and twenty-one-year-old son. She received her BS (1974) and MS (1976) in forestry and PhD (1978) in agronomy, specializing in plant genetics, from the University of Illinois (Champaign-Urbana). Her research is multidisciplinary and focuses on issues related to genetic diversity and urban forestry. Recently she has taught courses in urban forestry and genetic engineering. She developed and taught a course entitled "Women in Sciences: Shattering the Glass Ceiling" for the university honors program. She lives in Ellicott City, Maryland, with her husband, occasionally her two college-age children, two parrots, and numerous lizards and frogs. She can be contacted at mmcintos@umd.edu.

Marilyn Wilkey Merritt is an ethnographic linguist and education specialist who is a university lecturer and consultant. She is deeply committed to promoting human understanding and writes poetry and general essays as well as research articles. Trained at Northwestern University, Washington

University in St. Louis (MA, anthropology), and the University of Pennsylvania (PhD, linguistics), she also benefited from twelve postdoctoral years in India and Africa. Her interdisciplinary research considers social interaction, cognition, media, and individual creativity under conditions of social change. The mother of two children (four grandchildren), she lives with her social science husband of more than four decades in Arlington, Virginia. She can be reached at marilyn@merritt.to.

Emily Monosson is an environmental toxicologist, writer, consultant, and mother of two children aged eleven and thirteen. She received her BS from Union College, Schenectady, New York, in 1983 and her PhD in toxicology from Cornell University in 1988. Her focus as a toxicologist is the impact of emerging contaminants on human health and the environment, particularly on aquatic systems. She writes for local newspapers and journals (and for her own blog) as "The Neighborhood Toxicologist," reporting and interpreting current research on environmental contaminants. She is also involved in the development and stewardship of the Encyclopedia of Earth (http://www.eoearth.org). She lives in Montague, Massachusetts, with her husband and children. She can be contacted at emonosson@verizon.net.

Heidi Newberg is a tenured associate professor at Rensselaer Polytechnic Institute and the mother of four adorable children. Her husband, who has switched careers several times to follow her through her career, currently has a reentry position in bioinformatics research. She has contributed to the success of two high-profile astronomy projects: the Supernova Cosmology Project and the Sloan Digital Sky Survey. She has coauthored more than sixty refereed journal articles with a total of nine thousand citations and has also written several essays intended to help women succeed in physics careers, including "The Woman Physicist's Guide to Speaking," which was published in *Physics Today*. She can be contacted at heidi@rpi.edu.

Rachel Obbard is an experimental physicist with the British Antarctic Survey in Cambridge, England, where her current research is on the physics and chemistry of frost flowers and their contributions to the polar troposphere. She has two teenaged sons who are great companions, providing endless amusement and always challenging her to stay one step ahead of them. For respite she turns to her dog, Sarah. Rachel received her BSc in engineering physics from the Colorado School of Mines in 1985, an MSc in materials science and engineering from the University of New Hampshire in 2002, and a PhD in engineering (microstructural properties of ice) from Dartmouth College in 2006. She can be reached at rachel.w.obbard .th06@alum.dartmouth.org.

Catherine O'Riordan is a program director at the Consortium for Ocean Leadership and mother of two children aged nine and eleven. She received a BS in mechanical engineering in 1984 from Case Western Reserve University, and an MS in 1989 and PhD in 1994 in civil engineering (environmental fluid mechanics) from Stanford University. After six years conducting research on coastal processes in France, she left academia to become a program manager at the American Geophysical Union (AGU). She continues to engage other scientists in communicating their research results to decision makers and the public in her new program management position. In her spare time she trains for triathlons and endurance sports. She lives in Washington, DC, with her husband and children. She can be contacted at coriordan@oceanleadership.org.

Nanette J. Pazdernik is a cellular biologist, writer, adjunct professor, La Leche League leader, and mother to three children. She received her BA in biology from Lawrence University in Appleton, Wisconsin, in 1990 and her PhD in molecular, cellular, developmental biology and genetics from the University of Minnesota in 1996. She is currently working on a textbook entitled *Biotechnology: Applying the Genetic Revolution* for Elsevier, Inc. She is also teaching part-time as an adjunct professor at Southwestern Illinois University. Outside the science world, she helps mothers learn the art of breast-feeding as a La Leche League leader.

Devin Reese is currently a science teacher, naturalist, and mother of three children aged seven, five, and two. She received her BA from Harvard University in ethology and her PhD in integrative biology from the University of California, Berkeley, in 1996. She works for the National Science Resources Center as well as for a couple of community organizations in science education while serving as a primary caretaker for her primate troop. She lives with her husband and children in Alexandria, Virginia, and can be reached at devinreese@yahoo.com.

Marie Remiker is a pseudonym for the author of "Role Models: Out with the Old and in with the New."

Deborah Ross is a professor of biology, wife, and mother of a twenty-six-year old daughter. She received her PhD in environmental microbiology from Rutgers University in 1974. Her research interests in both industry and academia have focused on the biodegradation of pollutants by micro-organisms. She is now a full professor of biology at Indiana University-Purdue University Fort Wayne, where she teaches microbiology and introductory biology in both the biology department and the women's studies program. Currently her research focuses on molecular aspects of the biodegradation of polycyclic aromatic hydrocarbons. She can be reached at ross@ipfw.edu.

Christine Seroogy is an assistant professor at the University of Wisconsin and mother of two children aged five and seven years. She received her BS degree in microbiology in 1985 at the University of Wisconsin and MD degree at the University of Minnesota in 1993. Her training path included two years at Genentech, Inc. (south San Francisco) as a research technician before medical school; internship in pediatrics at the University of California-San Diego; residency in pediatrics at the Children's Hospital, Harvard University, Boston; fellowship in allergy/immunology at the University of California-San Francisco; and postdoctoral research at Stanford University. Her current position involves basic bench research in the area of cellular immunology and patient care in pediatric allergy/immunology/rheumatology. She has the privilege of training graduate and medical students as well as MDs in residency and fellowship. She lives in Madison, Wisconsin, with her two lovely children and can be reached at cmseroogy@wisc.edu.

Marguerite Toscano, mother of a ten-year-old daughter, settled on geology after serious flirtations with chemistry, marine biology, and music. She received a BS in geology from Long Island University in 1982. Work on coastal Quaternary geologic studies for her MS (University of Delaware 1986) and research at the Maryland Geological Survey led to a PhD (University of South Florida 1996) on late Pleistocene climate and sea levels from fossil coral reefs in the Florida Keys. As a postdoc, she has used satellite data to document the correlation between high water temperatures and coral bleaching worldwide. She has been a research associate of the Smithsonian Institution for eight years where her current work deals with Quaternary coral reef geological history and sea-level change in the Caribbean region. She still manages to incorporate scientific writing, music, a Girl Scout troop, and various artistic endeavors into her busy schedule.

Gina D. Wesley-Hunt is a paleontologist and evolutionary biologist. She graduated from Northwestern University in 1995 with a BA in evolutionary biology and then spent two terrific years jumping around the country joining field projects on small mammals, peregrine falcons, and fungi. In 2003 she received her PhD in evolutionary biology from the University of Chicago. Her area of research is mammalian evolution. She and her husband have just celebrated the birth of a wonderful daughter.

Theresa M. Wizemann is the mother of two budding scientists, aged five and twelve. She holds a bachelor's degree in medical technology from Douglass College of Rutgers University (1987)and master's and doctoral degrees in microbiology and molecular genetics, awarded jointly by Rutgers University and the University of Medicine and Dentistry of New Jersey (1992 and 1994, respectively). She spent her early career leading a vaccine research team at MedImmune, Inc., a biotechnology company. Following a year on Capitol Hill as a congressional science fellow, she became a study director for the Institute of Medicine of the National Academies. She currently works for Merck & Co., Inc., as a director of worldwide regulatory affairs, serving as the liaison to regulatory agencies for several infectious disease products, following several years as director of public policy, covering science- and research-related policy issues.

Sofia Katerina Refetoff Zahed studies the influence of hormones on parental behavior in small primates. She is the mother of two children aged one and four and is pursuing a PhD in zoology at the University of Wisconsin. She received her undergraduate degree at Washington University in St. Louis. She also worked as an environmental and science educator in the Chicago area helping students in the public schools, in the public park programs, and through public access television to have a better understanding of the natural sciences. She now lives in Madison with her husband and children. She can be contacted at skzahed@wisc.edu.

Gayle Barbin Zydlewski is a fish biologist, wife, mother of a seven-year-old, and watershed council president. She received her BS in biology and marine biology from Southeastern Massachusetts University in 1990, her MS in zoology from the University of Rhode Island in 1992, and her PhD in oceanography from the University of Maine in 1996. She conducted a postdoc at a United States Geological Survey laboratory, where she met and married her husband. In 1999, when she was seven months pregnant, they moved to the West Coast to study Pacific salmonids. She studies the relationship between the environment and fish behavioral and physiological responses. Her career passion is translating fisheries research results into useful fisheries management applications for resource protection. She lives in Hampden, Maine, with her husband and son. She can be reached at gayle.zydlewski@umit.maine.edu.